DSP for
MATLAB™ and LabVIEW™
Volume III: Digital Filter Design

SYNTHESIS LECTURES ON SIGNAL PROCESSING

Editor
José Moura, Carnegie Mellon University

DSP for MATLAB™ and LabVIEW™ Volume III: Digital Filter Design
Forester W. Isen

ISBN: 978-3-031-01402-4   paperback
ISBN: 978-3-031-02530-3   ebook

DOI 10.1007/978-3-031-02530-3

A Publication in the Springer series
SYNTHESIS LECTURES ON SIGNAL PROCESSING

Lecture #6
Series Editor: José Moura, Carnegie Mellon University

Series ISSN
Synthesis Lectures on Signal Processing
Print 1932-1236   Electronic 1932-1694

DSP for MATLAB™ and LabVIEW™
Volume III: Digital Filter Design

Forester W. Isen

SYNTHESIS LECTURES ON SIGNAL PROCESSING #6

ABSTRACT

This book is Volume III of the series DSP for *MATLAB*™ *and LabVIEW*™. Volume III covers digital filter design, including the specific topics of FIR design via windowed-ideal-lowpass filter, FIR highpass, bandpass, and bandstop filter design from windowed-ideal lowpass filters, FIR design using the transition-band-optimized Frequency Sampling technique (implemented by Inverse-DFT or Cosine/Sine Summation Formulas), design of equiripple FIRs of all standard types including Hilbert Transformers and Differentiators via the Remez Exchange Algorithm, design of Butterworth, Chebyshev (Types I and II), and Elliptic analog prototype lowpass filters, conversion of analog lowpass prototype filters to highpass, bandpass, and bandstop filters, and conversion of analog filters to digital filters using the Impulse Invariance and Bilinear Transform techniques. Certain filter topologies specific to FIRs are also discussed, as are two simple FIR types, the Comb and Moving Average filters. The entire series consists of four volumes that collectively cover basic digital signal processing in a practical and accessible manner, but which nonetheless include all essential foundation mathematics. As the series title implies, the scripts (of which there are more than 200) described in the text and supplied in code form (available via the internet at http://www.morganclaypool.com/page/isen) will run on both MATLAB and LabVIEW. The text for all volumes contains many examples, and many useful computational scripts, augmented by demonstration scripts and LabVIEW Virtual Instruments (VIs) that can be run to illustrate various signal processing concepts graphically on the user's computer screen. Volume I consists of four chapters that collectively set forth a brief overview of the field of digital signal processing, useful signals and concepts (including convolution, recursion, difference equations, LTI systems, etc), conversion from the continuous to discrete domain and back (i.e., analog-to-digital and digital-to-analog conversion), aliasing, the Nyquist rate, normalized frequency, sample rate conversion, and Mu-law compression, and signal processing principles including correlation, the correlation sequence, the Real DFT, correlation by convolution, matched filtering, simple FIR filters, and simple IIR filters. Chapter 4 of Volume I, in particular, provides an intuitive or "first principle" understanding of how digital filtering and frequency transforms work. Volume II provides detailed coverage of discrete frequency transforms, including a brief overview of common frequency transforms, both discrete and continuous, followed by detailed treatments of the Discrete Time Fourier Transform (DTFT), the z-Transform (including definition and properties, the inverse z-transform, frequency response via z-transform, and alternate filter realization topologies including Direct Form, Direct Form Transposed, Cascade Form, Parallel Form, and Lattice Form), and the Discrete Fourier Transform (DFT) (including Discrete Fourier Series, the DFT-IDFT pair, DFT of common signals, bin width, sampling duration, and sample rate, the FFT, the Goertzel Algorithm, Linear, Periodic, and Circular convolution, DFT Leakage, and computation of the Inverse DFT). Volume IV, the culmination of the series, is an introductory treatment of LMS Adaptive Filtering and applications, and covers cost functions, performance surfaces, coefficient perturbation to estimate the gradient, the LMS algorithm, response of the LMS algorithm to narrow-band signals, and various topologies such as ANC (Active Noise Cancelling) or system modeling, Periodic Signal Removal/Prediction/Adaptive

Line Enhancement (ALE), Interference Cancellation, Echo Cancellation (with single- and dual-H topologies), and Inverse Filtering/Deconvolution/Equalization.

KEYWORDS

Higher-Level Terms:

FIR Design, Classical IIR Design, Windowed Ideal Lowpass, Frequency Sampling, Equiripple, Remez Exchange

Lower-Level Terms:

Comb, Moving Average, Linear Phase, Passband Ripple, Stopband Attenuation, Highpass, Bandpass, Bandstop, Notch, Hilbert Transformer, Differentiator, Inverse-DFT, Cosine/Sine Summation Formulas, Alternation Theorem, Direct Form, Cascade Form, Linear Phase Form, Cascaded Linear Phase Form, Frequency Sampling Form, Butterworth, Chebyshev Type-I, Chebyshev Type-II, Elliptic, Cauer

This volume is dedicated to

Virginia L. (Durham) (Isen) Bowles
Renee J. (Udelson) Isen

and to the memory of the following

Forester W. Isen, Sr. (1916-1978)
James Daniel Mudd (1912-1997)
Glenn Warren McWhorter (1932-2008)

Contents

Preface to Volume III

0.1 INTRODUCTION

The present volume is Volume III of the series *DSP for MATLAB*™ *and LabVIEW*™. The entire series consists of four volumes which collectively form a work of twelve chapters that cover basic digital signal processing in a practical and accessible manner, but which nonetheless include essential foundation mathematics. The text is well-illustrated with examples involving practical computation using m-code or MathScript (as m-code is usually referred to in LabVIEW-based literature), and LabVIEW VIs.

There is also an ample supply of exercises, which consist of a mixture of paper-and-pencil exercises for simple computations, and script-writing projects having various levels of difficulty, from simple, requiring perhaps ten minutes to accomplish, to challenging, requiring several hours to accomplish. As the series title implies, the scripts given in the text and supplied in code form (available via the internet at `http://www.morganclaypool.com/page/isen`) are suitable for use with both MATLAB (a product of The Mathworks, Inc.), and LabVIEW (a product of National Instruments, Inc.). Appendix A in each volume of the series describes the naming convention for the software written for the book as well as basics for using the software with MATLAB and LabVIEW.

0.2 THE FOUR VOLUMES OF THE SERIES

The present volume, Volume III of the series, is devoted to Digital Filter Design. It covers FIR and IIR design, including general principles of FIR design, the effects of windowing and filter length, characteristics of four types of linear phase FIR, Comb and MA filters, Windowed Ideal Lowpass filter design, Frequency Sampling design with optimized transition band coefficients, Equiripple FIR design, and Classical IIR design.

Volume I of the series, Fundamentals of Discrete Signal Processing, consists of four chapters. The first chapter gives a brief overview of the field of digital signal processing. This is followed by a chapter detailing many useful signals and concepts, including convolution, recursion, difference equations, etc. The third chapter covers conversion from the continuous to discrete domain and back (i.e., analog-to-digital and digital-to-analog conversion), aliasing, the Nyquist rate, normalized frequency, conversion from one sample rate to another, waveform generation at various sample rates from stored wave data, and Mu-law compression. The fourth and final chapter of Volume I introduces the reader to many important principles of signal processing, including correlation, the correlation sequence, the Real DFT, correlation by convolution, matched filtering, simple FIR filters, and simple IIR filters.

Volume II of the series is devoted to discrete frequency transforms. It begins with an overview of a number of well-known continuous domain and discrete domain transforms, and covers the DTFT (Discrete Time Fourier Transform), the DFT (Discrete Fourier Transform), Fast Fourier Transform (FFT), and the z-Transform in detail. Filter realizations (or topologies) are also covered, including Direct, Cascade, Parallel, and Lattice forms.

Volume IV of the series, LMS Adaptive Filtering, begins by explaining cost functions and performance surfaces, followed by the use of gradient search techniques using coefficient perturbation, finally reaching the elegant and computationally efficient Least Mean Square (LMS) coefficient update algorithm. The issues of stability, convergence speed, and narrow-bandwidth signals are covered in a practical manner, with many illustrative scripts. In the second chapter of the volume, use of LMS adaptive filtering in various filtering applications and topologies is explored, including Active Noise Cancellation (ANC),system or plant modeling, periodic component elimination, Adaptive Line Enhancement (ADE), interference cancellation, echo cancellation, and equalization/deconvolution.

0.3   ORIGIN AND EVOLUTION OF THE SERIES

The manuscript from which the present series of four books has been made began with an idea to provide a basic course for intellectual property specialists and engineers that would provide more explanation and illustration of the subject matter than that found in conventional academic books. The idea to provide an accessible basic course in digital signal processing began in the mid-to-late 1990's when I was introduced to MATLAB by Dan Hunter, whose graduate school days occurred after the advent of both MATLAB and LabVIEW (mine did not). About the time I was seriously exploring the use of MATLAB to update my own knowledge of signal processing, Dr. Jeffrey Gluck began giving an in-house course at the agency on the topics of convolutional coding, trellis coding, etc., thus inspiring me to do likewise in the basics of DSP, a topic more in-tune to the needs of the unit I was supervising at the time. Two short courses were taught at the agency in 1999 and 2000 by myself and several others, including Dr. Hal Zintel, David Knepper, and Dr. Pinchus Laufer. In these courses we stressed audio and speech topics in addition to basic signal processing concepts. Thanks to The Mathworks, Inc., we were able to teach the in-house course with MATLAB on individual computers, and thanks to Jim Dwyer at the agency, we were able to acquire several server-based concurrent-usage MATLAB licenses, permitting anyone at the agency to have access to MATLAB. Some time after this, I decided to develop a complete course in book form, the previous courses having consisted of an ad hoc pastiche of topics presented in summary form on slides, augmented with visual presentations generated by custom-written scripts for MATLAB. An early draft of the book was kindly reviewed by Motorola Patent Attorney Sylvia Y. Chen, which encouraged me to contact Tom Robbins at Prentice-Hall concerning possible publication. By 2005, Tom was involved in starting a publishing operation at National Instruments, Inc., and introduced me to LabVIEW with the idea of possibly crafting a book on DSP to be compatible with LabVIEW. After review of a manuscript draft by a panel of three in early 2006, it was suggested that all essential foundation mathematics be included so the book would have both academic and professional appeal. Fortunately, I had long since retired from the agency and was able to devote the considerable amount of time needed for such a project. The result is a book suitable for use in both academic and professional settings, as it includes essential mathematical formulas and concepts as well as simple or "first principle" explanations that help give the reader a gentler entry into the more conventional mathematical treatment.

This double-pronged approach to the subject matter has, of course, resulted in a book of considerable length. Accordingly, it has been broken into four modules or volumes (described above) that together form a comprehensive course, but which may be used individually by readers who are not in need of a complete course.

Many thanks go not only to all those mentioned above, but to Joel Claypool of Morgan&Claypool, Dr. C.L.Tondo and his troops, and, no doubt, many others behind the scenes whose names I have never heard, for making possible the publication of this series of books.

Forester W. Isen
December 2008

CHAPTER 1

Principles of FIR Design

1.1 OVERVIEW

1.1.1 IN PREVIOUS VOLUMES

The previous volumes of the series are Volume I, *Fundamentals of Discrete Signal Processing*, and Volume II, *Discrete Frequency Transforms*. Volume I covers DSP fundamentals such as basic signals and LTI systems, difference equations, sampling, the Nyquist rate, normalized frequency, correlation, convolution, the real DFT, matched filtering, and basic IIR and FIR filters. Volume II covers the important discrete frequency transforms, namely, the Discrete Time Fourier Transform (DTFT), the Discrete Fourier Transform (DFT), and the z-transform.

With respect to digital filter design, Volumes I and II collectively provide necessary prerequisite knowledge of (1) digital filter topology and signal flow for both FIRs and IIRs; (2) how digital filters work to select and reject frequencies in a signal, and (3) how to evaluate the frequency-response of digital filters.

1.1.2 IN THIS VOLUME

In this volume, Volume III of the series, we take up digital filter design, which addresses the question of precisely how to design filters (i.e., compute an appropriate set of filter coefficients) that meet very specific design requirements, which include band limits, maximum passband ripple, minimum stopband attenuation, steepness of roll-off, etc. There are a number of design approaches for digital filters; they require knowledge of the DFT, the DTFT, the Laplace transform (for classical IIR filters), and the z-transform for complete understanding. How each of these transforms is relevant will become apparent as we move through this and the following two chapters.

1.1.3 IN THIS CHAPTER

In this chapter we acquire additional information and tools specific to the FIR, including the effects of filter length and windowing, the relationship between impulse response symmetry and phase linearity, and the frequency response of linear-phase FIRs. We then present brief discussions of two useful FIRs that are simple to design, the Comb Filter and the Moving Average Filter. These two filters have their own distinctive uses. The Comb filter is often useful for eliminating harmonically related spectral components in a signal, and the Moving Average filter is useful for signal averaging to emphasize a coherent signal in noise. The Comb Filter, additionally, is a processing component in the Frequency Sampling Form of FIR realization, which is covered at the end of the chapter along with other filter realizations particular to the FIR.

1.2   SOFTWARE FOR USE WITH THIS BOOK

The software files needed for use with this book (consisting of m-code (.m) files, VI files (.vi), and related support files) are available for download from the following website:

http://www.morganclaypool.com/page/isen

The entire software package should be stored in a single folder on the user's computer, and the full file name of the folder must be placed on the MATLAB or LabVIEW search path in accordance with the instructions provided by the respective software vendor (in case you have encountered this notice before, which is repeated for convenience in each chapter of the book, the software download only needs to be done once, as files for the entire series of four volumes are all contained in the one downloadable folder). See Appendix A for more information.

1.3   CHARACTERISTICS OF FIR FILTERS

- The impulse response of an FIR, if made symmetric or anti-symmetric, will yield a linear phase function, the result being that signals passing through the filter do not have their phases dispersed.

- Arbitrarily steep roll-offs may be had by making the impulse response correspondingly long. The cost is in computation, since linear convolution in the time domain of two sequences of length N requires about N^2 multiplications. This can often be alleviated by the use of frequency domain techniques.

- Arbitrary pass characteristics (i.e., other than standard lowpass, highpass, notch, or bandpass) may readily be generated.

- FIRs are inherently stable, meaning that a finite-valued input signal will never lead to an unbounded output signal.

1.4   EFFECT OF FILTER LENGTH

To discover several basic FIR principles, let's experiment with the length of a simple lowpass filter having impulse response $[ones(1, N)]$. By increasing N, the number of samples in the filter, we can observe the effect on frequency response, especially on steepness of roll-off. Figure 1.1 shows three such filters of increasing length in subplots (a), (c), and (e), and their frequency responses in plots (b), (d), and (f), respectively.

From Fig. 1.1 we can readily deduce that a filter can be made more selective or have a steeper roll-off by increasing its length. This may be explained by noting that as the filter length is increased, there are more distinct integral-valued frequencies (or correlators) between 0 and π radians (normalized frequencies of 0 and 1.0), that serve as potential correlation frequencies. From our studies of orthogonality (found in Volume I of the series; see the Preface of this volume for

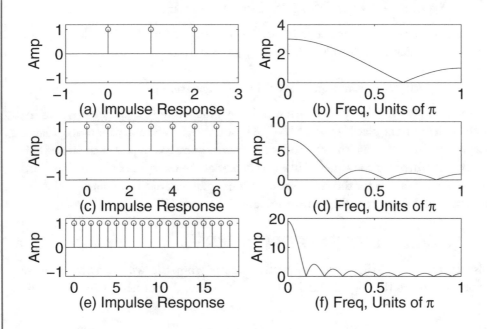

Figure 1.1: (a) 3-pt Rectangular Impulse Response; (b) Magnitude of frequency response of signal at (a); (c) 7-pt Rectangular Impulse Response; (d) Magnitude of frequency response of signal at (c); (d) 19-pt Rectangular Impulse Response; (e) Magnitude of frequency response of signal at (e).

information on Volume I), we know that if no correlator at a given frequency is in fact present in the impulse response, the filter's response at that precise frequency must be zero, assuming the presence of at least one other integral-valued frequency component (i.e., correlator) in the impulse response. In the case of the simple lowpass filter consisting of two or more samples valued at 1.0, only DC is present, so the frequency response must go to zero at each distinct correlator frequency other than zero. For the length-3 filter $[1,1,1]$, the only potential correlators are at normalized frequencies 0 and 0.6667 ($[0,1]/1.5$), and we can see in Fig. 1.1, subplot (b), that the frequency response goes to zero at frequency 0.6667. The length-7 filter shown in Fig. 1.1, subplot (c), has potential correlators at normalized frequencies of $[0:1:3]/(3.5) = [0, 0.2857, 0.5714, 0.8571]$. We can see from Fig. 1.1, subplot (d) that it indeed has a frequency response of zero at normalized frequencies of $[0.2857, 0.5714, 0.8571]$, as expected.

1.5   EFFECT OF WINDOWING

It can be seen in Fig. 1.1 that longer impulse responses, although more frequency selective, suffer from leakage or scalloping in the frequency response. Fortunately, this can be alleviated by using the same kind of windows on the filter impulse response that we used on signals prior to taking the

DFT. Figure 1.2 shows the same three impulse responses after smoothing with a *hamming* window, and the corresponding frequency responses. The improvement is dramatic.

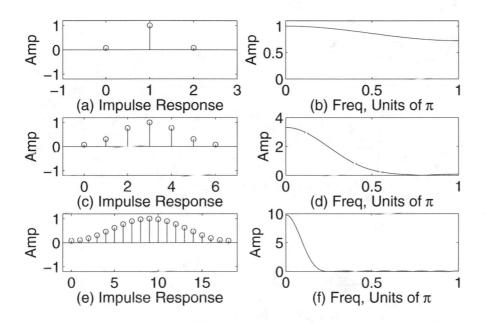

Figure 1.2: (a) 3-pt Hamming-windowed Impulse Response; (b) Magnitude of frequency response of signal at (a); (c) 7-pt Hamming-windowed Impulse Response; (d) Magnitude of frequency response of signal at (c); (d) 19-pt Hamming-windowed Impulse Response; (e) Magnitude of frequency response of signal at (e).

In Fig. 1.3, plot (a), a multi-cycle cosine (inherently rectangularly-windowed) has been used as the impulse response. As can be seen in plot (c), the expected scalloped response results. Figure 1.3, plot (d), shows the result when a *hamming* window is applied to the impulse response (plot (b)). Note that the steepness of roll-off of the main lobe of the response is decreased with use of the *hamming* window, compared to the rectangular window.

You should notice that the rectangular window result has a main lobe width narrower than that of the *hamming* window example, as can be readily seen, but the rectangular window's side lobe amplitude is very high (i.e., the stopband attenuation is poor), making the rectangular window unacceptable for most applications.

General Rule: For a given filter length, the greater the stopband attenuation, the shallower the main lobe roll-off must be. Stated conversely, for a given filter length, the steeper the main (or central) lobe roll-off, the poorer will be the ultimate stopband attenuation. For a given stopband attenuation, the roll-off may be improved by increasing the filter length.

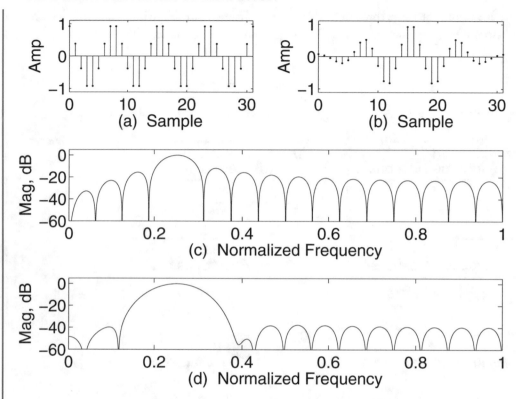

Figure 1.3: (a) Impulse response, a cosine, inherently windowed with a Rectangular window; (b) Same cosine as (a), multiplied by a Hamming window; (c) Magnitude of frequency response in dB of sequence at (a); (d) Magnitude of frequency response in dB of sequence at (b).

1.6   LINEAR PHASE

- FIRs can be made to have a **Linear Phase** response, which means that a graph of the phase shift imparted by the filter versus frequency is a straight line, or at least piece-wise linear.

- With linear phase shift, all frequencies remain in phase, and therefore the filter acts like a bulk delay line in which all frequencies remain in time alignment. Such a filter is said to be **Non-Dispersive**.

To see why a simple delay line has a linear phase delay characteristic (and vice versa), imagine a signal simply going through a delay line that imparts a delay time of τ. A given frequency f_0 has a period of $1/f_0$ and is therefore delayed by $\tau/(1/f_0) = f_0 \cdot \tau$ cycles, which is just a linear function of frequency If f_0 doubles, for example, so does the number of cycles of delay, and so on.

Example 1.1.   Using a signal having components of 100 Hz, 200 Hz, 300 Hz, etc., show that a delay line which imparts a bulk delay of 0.01 second imparts a linear phase shift.

A 100 Hz signal has one hundred cycles in one second, or one cycle in 0.01 second, so clearly it is delayed one cycle, or 2π radians. The 200 Hz component is delayed by two cycles, since two of its cycles occur in the 0.01 second delay time. Hence, it is delayed by 4π radians, and so forth. You can see that if all the components are in phase when they go into the delay line, they will still be in phase when they come out, since the 100 Hz component will be delayed exactly one cycle, the 200 Hz component exactly two cycles, and so on, so that on exiting, they are still all beginning their cycles together, i.e., in phase with each other.

- Modern communication signals often involve pulses or square-wave-like shapes, and detection of the signal often depends on its time domain shape, not its frequency components *per se*, so it is beneficial when passing such a signal through a filter (to remove high frequency noise or the like) that the filter not disperse the phases of the frequency components, since dispersal would cause the waveform to lose its shape and perhaps its detectability as well. In music, for example, too much phase dispersion can cause audible distortion, especially in fast transients (characteristic of percussion instruments, for example) that depend on proper phase alignment for their sharp definition in time.

- Linear phase filters always impart a constant delay to a signal equal to one-half the filter length.

Sometimes the filter length must become very large to achieve certain design criteria, and the delay time may become unacceptable. In such cases, it is possible to use an FIR with nonlinear phase which has an acceptable magnitude response, but a much decreased delay. On the other hand, when linear phase is not necessary, IIRs are often used since the computational burden is usually much less for magnitude responses similar to those of an FIR.

1.6.1   IMPULSE RESPONSE REQUIREMENT
- Making a linear phase FIR impulse response is not difficult. All that is required is that the impulse response coefficients be either symmetrical or anti-symmetrical about the middle of the impulse response.

For example, a length-seven symmetrical impulse response might look like this: $[a, b, c, d, c, b, a]$, whereas a length-six filter would be $[a, b, c, c, b, a]$. The first and last coefficients are the same, the second and penultimate coefficients are the same, and so forth. An anti-symmetric impulse response of length seven might be $[a, b, c, 0, -c, -b, -a]$, whereas a length-six anti-symmetrical would be $[a, b, c, -c, -b, -a]]$. In this case, the first and last coefficients have opposite signs, and so on.

Figure 1.4 compares two impulse responses, their frequency content, and their phase responses. At (a), one cycle of a nonsymmetrical cosine has its frequency response (via DTFT) shown at (c), and it phase response at (e). Note that the phase response is slightly nonlinear at low frequencies. At (b), the same one cycle cosine has been slightly adjusted to make it symmetrical, resulting in a perfectly piece-wise linear phase characteristic at (f), having very nearly the same magnitude response (at (d)) as the nonsymmetrical impulse response.

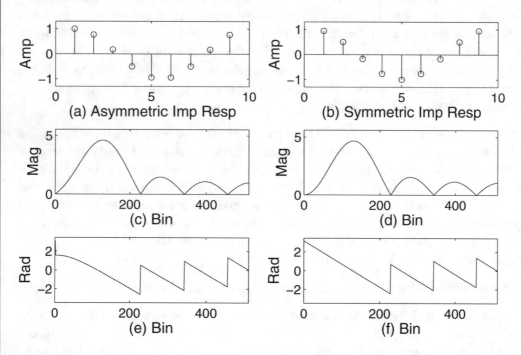

Figure 1.4: (a) Asymmetric cosine; (b) Symmetric cosine; (c) Magnitude of spectrum of signal at (a); (d) Magnitude of spectrum of signal at (b); (e) Phase response of signal at (a); (f) Phase response of signal at (b).

A much more egregious example, shown in Fig. 1.5, is had by making the simple impulse response [1.5,0.5], which consists of two frequency correlators, DC and 1 cycle, weighted with 1 and 0.5, respectively, i.e., [1.5,0.5] = [1,1] + 0.5[1,-1]. The impulse response is clearly nonsymmetric, as is the phase response.

1.6.2   FOUR BASIC CATEGORIES OF FIR IMPULSE RESPONSE FOR LINEAR PHASE

Linear phase impulse responses must be either symmetric or anti-symmetric, and any impulse response must have either an even length (evenly divisible by two) or an odd length. There are distinct differences between even and odd length filters, as well as symmetric and anti-symmetric filters.

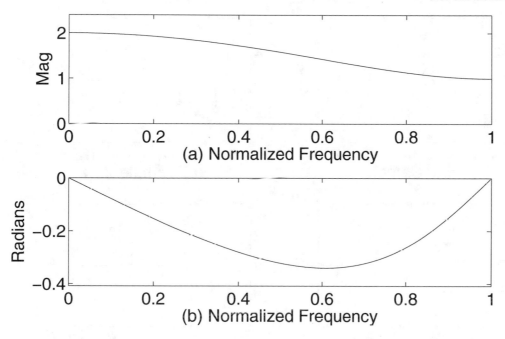

Figure 1.5: (a) Magnitude of the DTFT of the impulse response [1.5 0.5]; (b) Phase response of the same impulse response.

Considering both symmetry and length, there are four basic FIR types, which are illustrated in Fig. 1.6.

Type I, Symmetric, Odd Length
This is perhaps the most used of linear phase FIR filter types because it is suitable for lowpass, highpass, bandpass, and bandstop filters. Correlator basis functions are cosines, meaning that the impulse response is generated by weighting and summing cosines of different frequencies.

Type II: Symmetric, Even Length
This filter type cannot be used as a highpass or bandstop filter since the cosine, at the Nyquist limit (necessary in the impulse response for highpass or bandstop filters), cannot be symmetrical in an even length. Lowpass and bandpass filters are possible.

Type III: Anti-Symmetric, Odd Length
The correlator basis functions are sines. The basis functions are identically zero at DC and the Nyquist limit, so this type of filter cannot be used for lowpass or highpass characteristics. Bandpass filters are possible, but the primary use is in designing Hilbert transformers and differentiators. Hilbert transformers shift the phase of all frequencies in a signal by 90°, or $\pi/2$ radians. This is useful

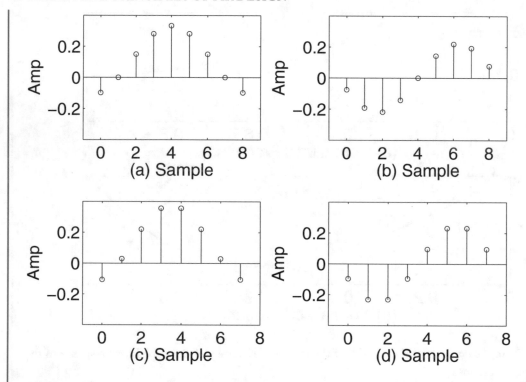

Figure 1.6: (a) Symmetric, odd-length FIR; (b) Anti-symmetric, odd-length FIR; (c) Symmetric, even-length FIR; (d) Anti-symmetric, even-length FIR.

in certain communications applications, such as generating single sideband signals, demodulating quadrature modulated signals, and so forth.

Type IV: Anti-Symmetric, Even Length
Like the Type III filter, this filter uses sines as the basis correlating functions. It is not suitable for a lowpass characteristic, but is suitable for Hilbert transformers and differentiators.

1.6.3   ZERO LOCATION IN LINEAR PHASE FILTERS

A linear phase FIR's zeros conform to certain requirements. Zeros having magnitude other than 1.0 (i.e., not on the unit circle) must always be matched with a zero having the same frequency but the reciprocal magnitude. Any complex zero, of course, must also be matched with its complex conjugate to ensure real-only coefficients. The following possibilities therefore exist for zeros for linear phase FIRs:

- Zeros at 1 and -1 may exist singly since they are real and have magnitude of 1.0.

- Any zero on the real axis having magnitude other than one must be matched by a real zero having the same frequency (0 or π radians) and the reciprocal magnitude.

- Any nonreal zero on the Unit Circle must be matched with its complex conjugate. Such zeros, then, can be characterized as existing in pairs.

- Any zero that is complex and has a magnitude other than 1.0 must be matched by a zero at the same frequency and reciprocal magnitude, and both of these zeros must be matched by their complex conjugates. Therefore, this type of zero comes in sets of four, called **Quads**.

The LabVIEW VI

$$DemoDragZerosZxformVI$$

allows you to drag a single zero in the z-plane, and have the single zero, or a complex conjugate pair of zeros, or a quad of zeros, computed and displayed, along with the impulse response and the z-transform magnitude and phase responses on a real-time basis. When operating in complex conjugate pair mode, if the imaginary part is zero, the mode devolves to single-zero mode. When in quad-zero mode, if the magnitude of the cursor zero is set to 1.0, and the imaginary part is not zero, then the mode devolves to complex conjugate mode; if the imaginary part is zero and the magnitude is 1.0, then mode devolves to single-zero, but if the magnitude is not 1.0, then two zeros are created, being equal to the cursor zero and its reciprocal. In this way, the VI always uses the minimum number of zeros necessary to construct a linear-phase FIR. Figure 1.7 shows an example of the VI with a quad of zeros–note the characteristic linear phase characteristic and all-real impulse response.

A script for use with MATLAB that performs functions similar to that of the VI above is

$$ML\_DragZeros$$

This script, when called, opens a GUI that allows you to select a single zero, a complex conjugate pair of zeros, or a quad of zeros (a complex conjugate pair and their reciprocals). The magnitude and phase of the z-transform as well as the real and imaginary parts of the equivalent impulse response are dynamically plotted as you move the cursor in the z-plane. This script, unlike the VI above, does not devolve to use of one or two zeros in certain cases, nor is an attempt made to maintain phase linearity. In the quad mode, four zeros are always used to create the FIR transfer function, and in this case, of course, phase is always linear. In the complex conjugate mode, phase is only linear when the magnitude of the zeros is 1.0, and in the single zero mode, phase is only linear when the zero is real with magnitude equal to 1.0. The reader should verify these observations when running the script.

Example 1.2.   Linearize the phase response of an FIR that has the transfer function

$$H(z) = 1 - 0.81z^{-2}$$

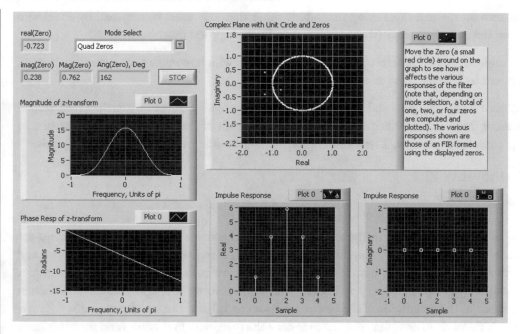

Figure 1.7: A VI that allows study of the effect of zero location and aggregation (i.e., single zero, complex conjugate pairs, or quads) on frequency, phase, and impulse responses of an FIR constructed with the computed and displayed zero(s).

First, we compute the values of its zeros as 0.9 and -0.9. The we supply two more zeros, the first at frequency zero (DC) with magnitude 1/0.9, and the second at the Nyquist frequency, with magnitude 1/0.9. To get the new transfer function, we convolve the factors [1,-0.9]. [1,-1/0.9], [1,0.9], and [1,1/0.9] and get

$$H(z) = 1 + 0z^{-1} - 2.0446z^{-2} + 0z^{-3} + z^{-4}$$

which gives the impulse response as [1,0,-2.0446,0,1], which is a linear phase bandpass filter with the passband centered at a normalized frequency of 0.5 (one-quarter of the sampling frequency).

You can verify this by making either of the following calls:

LVxFreqRespViaZxform([1, 0, -2.0446, 0, 1],1024)

or

y = abs(fft([1, 0, -2.0446, 0, 1],1024)); plot(y(1,1:512))

Example 1.3.   It is known that a certain linear phase FIR has zeros at $0.9j$, $(\sqrt{2}/2)(1 + j)$, and -1.0. Give the entire set of zeros.

The zero at 0.9j gives rise to another zero at its reciprocal magnitude, and two more zeros that are the complex conjugates of the first two zeros. The zero at $(\sqrt{2}/2)(1 + j)$ is on the Unit Circle, and thus gives rise only to its own complex conjugate. The zero at -1 is both real and of magnitude 1.0 and thus exists by itself. The entire list is therefore 0.9j, -0.9j, 1.11j, -1.11j, 0.707(1 + j), 0.707(1 - j), and -1.0.

Example 1.4.   Obtain the FIR coefficients corresponding to the zeros in the above example.

We make the call

$$a = 0.9^*j; b = \exp(j^*pi/4); Imp = poly([a,-a,(1/a),(1/-a),b,b\char`\^-1,-1.0])$$

which returns the coefficients as Imp.

1.7   LINEAR PHASE FIR FREQUENCY CONTENT AND RESPONSE

Since linear phase FIRs conform to specific forms for the impulse response, the frequency response can be written for each of the four FIR types as an expression involving cosines or sines. The **Impulse Response** for **Type I or II filters** conforms to the rule that

$$h[n] = h[L - n - 1]$$

where the impulse response length is L and n = 0:1:L − 1.

The **Frequency Response** for **Type I and II filters** conforms to the general form

$$H(\omega) = H_r(\omega)e^{-j\omega M}$$

where $M = (L − 1)/2$, and $Hr(\omega)$ is a real function that can be positive or negative and is therefore called the **Amplitude Response,** while the complex exponential represents a linear phase factor.

For the Type I filter (odd length), the amplitude response is given as

$$H_r(\omega) = h[M] + 2 \sum_{n=1}^{M} h[M - n] \cos(\omega n) \tag{1.1}$$

which is equivalent to

$$H_r(\omega) = h[M] + 2 \sum_{n=0}^{M-1} h[n] \cos(\omega[M - n]) \tag{1.2}$$

Example 1.5.   Consider the Type I linear phase filter having the impulse response [a, b, c, b, a]; show that Eq. (1.1) is equivalent to the DTFT of the impulse response.

The DTFT is defined as

$$X(e^{j\omega}) = \sum_{n=-\infty}^{\infty} x[n]e^{-j\omega n}$$

from which we get

$$X(e^{j\omega}) = a + be^{-j\omega} + ce^{-j\omega 2} + be^{-j\omega 3} + ae^{-j\omega 4}$$

which reduces to

$$X(e^{j\omega}) = e^{-j\omega 2}(ae^{j\omega 2} + be^{j\omega} + c + be^{-j\omega} + ae^{-j\omega 2})$$

and finally

$$X(e^{j\omega}) = e^{-j\omega 2}(c + 2b\cos(\omega) + 2a\cos(2\omega))$$

which can be written as

$$X(e^{j\omega}) = e^{-j\omega M}[x[M]\cos(0\omega) + 2(x[M-1]\cos(1\omega) + x[M-2]\cos(2\omega))]$$

where $M = (L-1)/2$ which conforms to Eq. (1.1).

For the Type II (even length) filter $H_r(\omega)$ is given as

$$H_r(\omega) = 2\sum_{n=0}^{L/2-1} h[n]\cos(\omega[M-n]) \tag{1.3}$$

which is equivalent to

$$H_r(\omega) = 2\sum_{n=1}^{L/2} h[\frac{L}{2} - n]\cos(\omega[n - \frac{1}{2}]) \tag{1.4}$$

The **Impulse Response** for **Type III and IV filters** conforms to the rule

$$h[n] = -h[L - n - 1]$$

and the **Frequency Response** for **Type III and IV filters** conforms to the general form

$$H(\omega) = jH_r(\omega)e^{-j\omega M} = H_r(\omega)e^{j(\pi/2 - \omega M)}$$

where $H_r(\omega)$, for Type III (odd length) is given as

$$H_r(\omega) = 2 \sum_{n=0}^{M-1} h[n] \sin(\omega[M-n]) \tag{1.5}$$

and for Type IV (even length) $H_r(\omega)$ is

$$H_r(\omega) = 2 \sum_{n=0}^{L/2-1} h[n] \sin(\omega[M-n]) \tag{1.6}$$

Example 1.6. Compute and plot the amplitude and phase responses for the Type I linear phase filter whose impulse response is $[1, 0, 1]$. Contrast the amplitude and corresponding phase plots to magnitude and phase plots obtained using the DTFT.

For the amplitude response we get $L = 3$, $M = 1$, and the formula

$$H_r(\omega) = h[M] + 2 \sum_{n=0}^{M-1} h[n] \cos(\omega[M-n])$$

becomes

$$H_r(\omega) = 0 + 2 \sum_{n=0}^{0} h[0] \cos(\omega[1-0]) = 2\cos(\omega)$$

and the phase is $e^{-j\omega}$. We use the following code to evaluate and plot the result at a large number of frequencies between $-\pi$ and π:

```
function LVFrPhRImp101
incF = 2*pi/1024; argF = -pi+incF:incF:pi;
Hr = 2*cos(argF); Hph = -argF;
figure(8); subplot(221); plot(argF/pi, Hr)
subplot(222); plot(argF/pi,Hph)
H = fft([1 0 1],1024); Hdtft = fftshift(H); subplot(223);
xvec = [-511:1:512]/512; plot(xvec, abs(Hdtft));
subplot(224); plot(xvec, unwrap(angle(Hdtft)))
```

Figure 1.8 shows the result from running the above code. Note that the amplitude and corresponding phase functions are linear and continuous, whereas the magnitude and corresponding phase functions are not.

Example 1.7. Compute and plot the amplitude and phase responses and magnitude and phase responses for the Type II filter whose impulse response is $[1, 1, 1, 1]$.

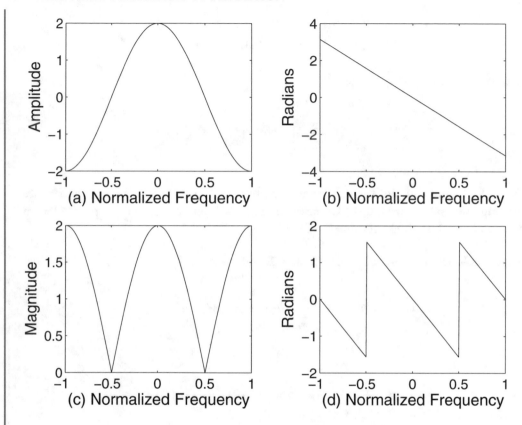

Figure 1.8: (a) Amplitude response of Type I linear phase filter whose impulse response is [1,0,1]; (b) Phase response or function of same; (c) Magnitude response of same impulse response; (d) Phase response of same.

We use the Eq. (1.3), with $L = 4$ and $M = 3/2$, we get

$$H_r(\omega) = 2 \sum_{n=0}^{1} h[n] \cos(\omega[\frac{3}{2} - n]) = 2(\cos(\frac{3}{2}\omega) + \cos(\frac{1}{2}\omega))$$

Using code similar to that given for the previous example, we get Fig. 1.9.

Example 1.8.   Compute and display the amplitude and phase responses and magnitude and phase responses for the Type III filter having impulse response [1,0,-1].

Using $M = 1$ and Eq. (1.5) we get

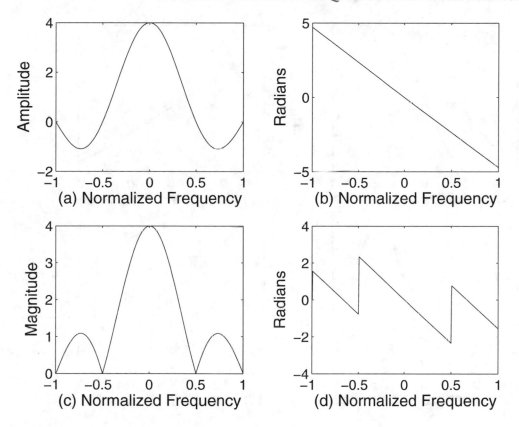

Figure 1.9: (a) Amplitude response of Type II linear phase filter whose impulse response is [1,1,1,1]; (b) Phase response or function of same; (c) Magnitude response of same impulse response; (d) Phase response of same.

$$H_r(\omega) = 2 \sum_{n=0}^{0} h[0] \sin(\omega[1 - 0]) = 2 \sin(\omega) \tag{1.7}$$

and the corresponding frequency and phase responses are shown in Fig. 1.10.

Example 1.9. Compute and plot the amplitude and phase responses and magnitude and phase responses for the Type IV filter whose impulse response is [1, 1,-1,-1].

Using Eq. (1.6) with $M = 1.5$, we get

$$H_r(\omega) = 2 \sum_{n=0}^{1} h[n] \sin(\omega[\frac{3}{2} - n]) = 2(\sin(\frac{3}{2}\omega) + \sin(\frac{1}{2}\omega))$$

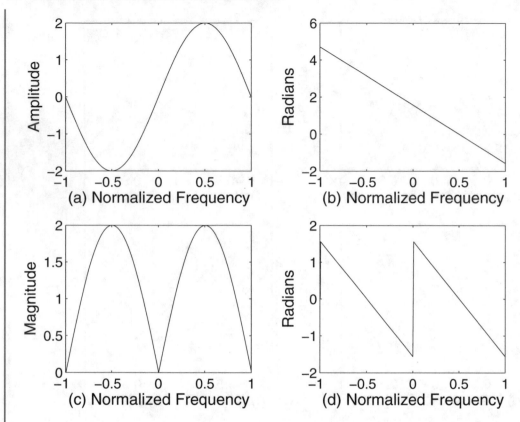

Figure 1.10: (a) Amplitude response of Type III linear phase filter whose impulse response is [1, 0, -1]; (b) Phase response or function of same; (c) Magnitude response of same impulse response; (d) Phase response of same.

Figure 1.11 shows the result.

1.8   DESIGN METHODS

1.8.1   BASIC SCHEME

- Designing frequency-selective filters consists of specifying **Passband(s)** (frequencies to be passed unattenuated), **Stopband(s)** (frequencies to be completely attenuated), and **Transition Band(s)**, containing frequencies lying between the passband(s) and stopband(s) which may (with some constraints) have whatever amplitudes are necessary to help optimize the responses in the passband(s) and stopband(s). Thus, the entire possible frequency range from 0 to the Nyquist limit is broken into one or more passbands, stopbands, and transition bands.

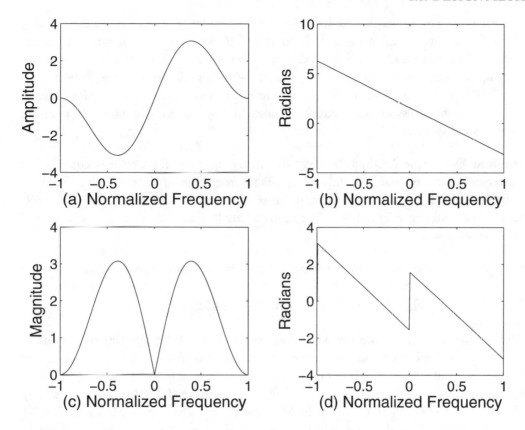

Figure 1.11: (a) Amplitude response of Type IV linear phase filter whose impulse response is [1,1,-1,-1]; (b) Phase response or function of same; (c) Magnitude response of same impulse response; (d) Phase response of same.

- All linear-phase FIRs are composed of (or synthesized by summing) symmetrical cosines or sines (i.e., correlators) having frequencies conforming to either of two orthogonal frequency schemes, namely, whole integer frequencies such as 0, 1, 2, etc., or odd multiples of half-cycles, yielding frequencies of 0.5, 1.5, 2.5, etc. The net frequency response is the superposition of the individual frequency responses contributed by each correlator.

1.8.2   THREE DESIGN METHODS

- One approach, called the **Window Method**, is to generate a truncated "ideal" lowpass filter and apply a window to the impulse response to achieve a certain desired stopband attenuation or reduction of passband ripple. Filter length is adjusted as necessary to achieve desired roll-off rate. Other filters such as highpass, bandpass, and notch can be generated starting with one or more lowpass filters.

- In the **Frequency Sampling Method**, the desired magnitudes of filter response at a plurality of DFT frequencies (i.e., frequencies defined as $2\pi k/L$ with $k = 0{:}1{:}L\text{-}1$, for example) are specified, a linear phase factor is imparted, and the inverse DFT is computed to obtain the filter's impulse response. A variant on the Frequency Sampling Method is to use, instead of the inverse DFT, simple cosine or sine summation formulas that construct an impulse response as the superposition of symmetrical cosines or sines, having frequencies conforming to one of two orthogonal systems.

- **Optimized Equiripple Method**. In this method, the approximation error is equalized in the passband(s) and stopband(s), with the maximum magnitude of error in each being user-specifiable. This method offers the greatest degree of user-control of the three methods discussed herein, and generally results in the shortest length filter that can meet a given set of design specifications.

- **A detailed discussion of these three design methods is found in the next chapter.**

1.8.3   THE COMB AND MOVING AVERAGE FILTERS

The Comb Filter

The Comb filter has only two nonzero values in its impulse response. A single delayed version of an input sequence is added to or subtracted from the undelayed version of the input sequence to form the output. For the case of a single sample of delay, the impulse response would be [1,1] when the delayed signal is added to the original, or [1,-1] when it is subtracted. For two samples of delay, it would be [1,0,1] or [1,0,-1], and so forth for different delays. The simple FIRs [1,1], [1,-1], [1,0,-1], and [1,0,1] are all comb filters which may also be characterized, respectively, as lowpass, highpass, bandpass, and notch filters. When the second non-zero value in the impulse response does not have unity magnitude, the null-depth does not go to zero-magnitude. Examples of such impulse responses would be [1, 0.9], [1, 0, -0.7], etc. In this book, this type of impulse response will generally be referred to as a Modified Comb Filter.

Comb filters are useful in certain types of applications. Suppose, for example, that you had an audio signal polluted with a 60 Hz fundamental wave with very high harmonic amplitudes extending into the 10^4 or higher frequency range. A comb filter is ideal for suppressing such a harmonic series, exhibiting economy and simplicity.

Figure 1.12 shows the frequency responses of two five-sample-delay comb filters, the first additive, and the second subtractive.

The script

```
function LVCombFilter(Tau)
% LVCombFilter(5)
ImpAdd = [1,zeros(1,Tau-1),1];
ImpSub = [1,zeros(1,Tau-1),-1];
DTFTLen = 1024; xplot = [1:1:DTFTLen/2+1];
```

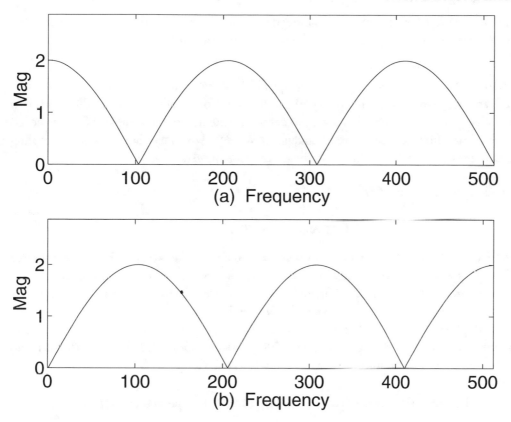

Figure 1.12: (a) Frequency response of the impulse response [1,0,0,0,0,1] (an additive comb filter having 5 samples of delay); (b) Frequency response of the impulse response [1,0,0,0,0,-1] (a subtractive comb filter having 5 samples of delay).

```
xvec= (xplot-1)/(DTFTLen/2); subplot(2,1,1);
yAdd = abs(fft(ImpAdd,DTFTLen));
plot(xvec,yAdd(xplot),'b'); ylabel(['Magnitude'])
xlabel(['(a) Normalized Frequency'])
axis([0 1 0 1.2*max(yAdd)])
subplot(2,1,2); ySub = abs(fft(ImpSub,DTFTLen));
plot(xvec,ySub(xplot),'b'); ylabel(['Magnitude'])
xlabel(['(b) Normalized Frequency'])
axis([0 1 0 1.2*max(ySub)])
```

affords experimentation with additive and subtractive comb filters; the variable *Tau* is the number of samples of delay. Figure 1.12 was generated by making the call **LVCombFilter(5)**.

Example 1.10.   Derive empirically, using the script *LVCombFilter(Tau)* with values of *Tau* such as 1, 2, 3, etc., an expression for the frequency of the first null for an additive comb filter having n samples of delay. Use the derived relationship to determine the number of samples of delay needed for an additive comb filter to have its first frequency null at 60 Hz when a signal having a sampling rate of 3000 Hz is convolved with the comb filter. Test your answer.

By making a succession of calls

$$LVCombFilter(Tau)$$

where Tau = 1, 2, 3, etc., and observing plot (a) of the resulting figure, we note that for one sample of delay (i.e., Tau = 1), the first null is at normalized frequency 1.0, for two samples delay, at 1/2 (0.5), for three samples delay, at 1/3 (0.333), and so forth, suggesting that the normalized frequency of the first null is at $1/Tau$ (this applies to an additive comb filter).

For a sampling rate of 3000 Hz, the Nyquist frequency is 1500 Hz, and therefore the normalized frequency desired for the first null is 60/1500 = 0.04, which is 1/25. We therefore need an additive comb filter with 25 samples of delay. A call which will verify this is:

FR = abs(fft([1,zeros(1,24),1],3000)); plot(0:1:180, FR(1,1:181))

Note that the first null is at 60 Hz, and the second one at 180 Hz.

The MA (Moving Average) Filter

The Moving Average filter is a single-correlator linear phase filter, the single correlator being at frequency zero (DC) The impulse response is that of a rectangle with a length of N samples, weighted by $1/N$:

$$(1/N) \cdot [1, 1....1]$$

The MA filter is useful for keeping a running average of the values of an input sequence over a certain length. An advantage of the MA impulse response is that it is possible to compute it recursively, eliminating a large amount of time-consuming convolution. To do this, let's first dispose of the scaling constant $1/N$ prior to discussing the recursion process. Note that we can either scale every input sample by $1/N$ before adding, or instead add up all samples and then scale just before delivering the sum as the next output. Let's assume the latter, so that all our operations prior to scaling by $1/N$ involve the raw, unscaled input samples.

The value we compute, then, prior to the final scaling, we will call a running sum S, which is just the output sequence generated as

$$S(n) = \sum_{i=n-N+1}^{n} x_i$$

and where N is the impulse response length, M is the length of the signal sequence x_i, which is defined as 0 for $i < 1$ or $i > M$; and n, for valid output (i.e., the impulse response saturated with input samples), runs from N to $(M - N + 1)$.

There is a very efficient way to compute the output sequence values of a running sum. Suppose that $S[8]$, which is the sum of input samples 1 through 8 (x_1 through x_8), has just been computed. To compute $S[9]$, which is the sum of x_2 through x_9, add x_9 to $S[8]$, and subtract x_1. Once $S[9]$ is in hand, of course, $S[10]$ can be computed by adding x_{10} and subtracting x_2, and so forth. Phrased mathematically, this would be:

$$S[n] = S[n - 1] + x_n - x_{n-N}$$

This simple recursion formula can greatly reduce computational overhead when N gets to be very large.

Since a Moving Average filter has an impulse response which consists of samples of a cosine of frequency zero, we would expect orthogonal behavior toward signals having an integral number of cycles in the length of the MA filter. For example, a 20-point MA impulse response will yield an output identically zero (during saturation of the filter, of course) for sinusoids having exactly one cycle, two cycles, three cycles, etc. up to ten cycles in a length of 20 samples. Figure 1.13 shows a 20-pt MA filter impulse response in plot (a), and its frequency response in plot (b).

In some situations, it is possible to enhance the signal-to-noise ratio (SNR) of a signal by using the MA filter. This requires that the noise be generally random or *incoherent*, and that the signal be coherent, or very predictable. The easiest situation is when the signal is all of one polarity. In this case, the signal, when averaged, builds up approximately proportionately to N, the number of samples added together, while the standard deviation of the noise builds up only by $\sqrt{N}$, thus increasing the SNR by $\sqrt{N}$.

Looking again at Fig. 1.13, we see in plot (c) what appears to be nothing but noise. In reality, it is a large amount of noise to which has been added five equally spaced signals, each consisting of a sequence of 20 samples valued at 1. Plot (d) of Fig. 1.13 shows five signal peaks that have been recovered from the noise by using the 20-pt moving average filter shown in the upper left plot. The best result occurs (on the average) when the MA filter length is exactly the same length as the coherent signal, since at that point, the maximum signal gain would have been obtained; averaging in more samples would only bring in additional noise without bringing in any additional signal.

Figure 1.14 shows the result when the MA filter is only 5 samples long. The results of the averaging in plot (d) show that a number of the signal pulses in the noise have not been well-recognized or emphasized. The results, however, vary with each running since the test signal is random noise, which is different for each trial run.

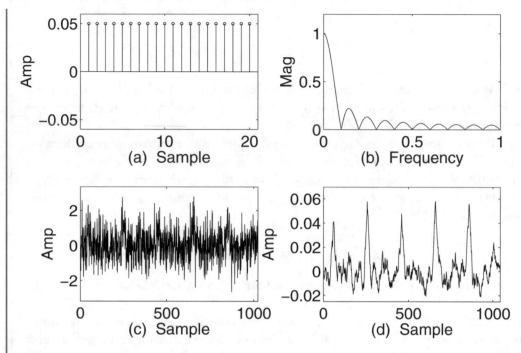

Figure 1.13: (a) 20-point Moving Average Impulse Response; (b) Normalized frequency response of impulse response in (a); (c) Test signal of noise with coherent (rectangular) signals embedded; (d) Convolution of test signal from (c) with impulse response from (a).

1.9   FIR REALIZATION

In Volume II of the series (see the Preface to this volume for information on Volume II), we explored the Direct, Cascade, Parallel, and Lattice Forms of realization for a generalized LTI system having both IIR and FIR components. In addition to these forms, there are several implementations that apply specifically to FIRs, including the Linear Phase FIR, which we have just introduced in this chapter. We begin with the simple Direct or Transversal form, proceed through Cascade and Linear Phase Forms, and finish with the Frequency Sampling Form, which is based on the idea of reconstructing the z-transform of an FIR from its samples, which we discussed in conjunction with the Discrete Fourier Series in Volume II.

1.9.1   DIRECT FORM

The Direct Form implementation of an FIR, illustrated in Fig. 1.15, is the simple transversal arrangement, in which the signal passes down a chain of delay elements, and the output of each delay element is weighted by a respective coefficient and all weighted outputs are summed to produce

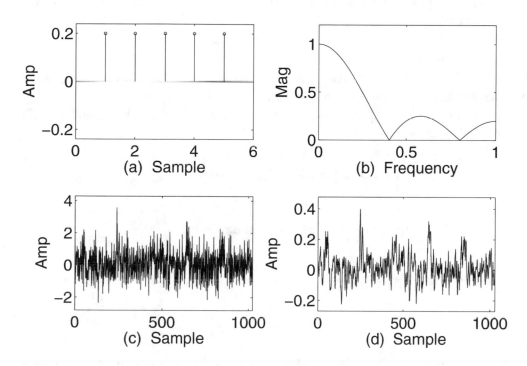

Figure 1.14: (a) 5-point Moving Average Impulse Response; (b) Normalized frequency response of impulse response in (a); (c) Test signal of noise with coherent (rectangular) signals embedded; (d) Convolution of test signal from (c) with impulse response from (a).

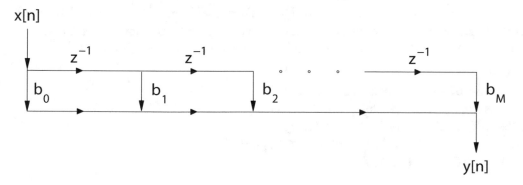

Figure 1.15: An FIR implemented in Direct Form. This arrangement is often called a Transversal filter since the signal moves across the filter. Note that the input signal $x[n]$ and the outputs of all delay elements are scaled by respective coefficients b_i and then summed to generate the output $y[n]$.

the output. In this arrangement, the filter signal flow diagram can be constructed directly from the z-transform of the FIR.

1.9.2   CASCADE FORM

By collecting the FIR's zeros in complex conjugate pairs, second order real sections can be made and cascaded to implement the filter, as shown in Fig. 1.16.

The coefficients can be computed using the same scripts we used for the generalized LTI Cascade Form, namely, the scripts

$$[Bc, Ac, Gain] = LVDirToCascade(b, a)$$

$$[b, a, k] = LVCas2Dir(Bc, Ac, Gain)$$

$$[y] = LVCascadeFormFilter(Bc, Ac, Gain, x)$$

Example 1.11.   For the Linear Phase filter having b = $fir1(6, 0.5)$, compute the Cascade Form coefficients, and filter a linear chip using the Direct and Cascade Form coefficients and compare the results.

The following m-code generates a set of Direct Form coefficients for a lowpass FIR, then computes the Cascade Form coefficients, filters a test chirp using both forms, plots the results (shown in Fig. 1.17), and then converts the Cascade Form coefficients back to Direct Form.

```
[b] = fir1(6,0.5);
[Bc,Ac,Gain] = LVDirToCascade(b,1)
x = chirp([0:1/999:1],0,1,500); y = filter(b,1,x);
[y2] = LVCascadeFormFilter(Bc,Ac,Gain,x);
figure(6); subplot(211); plot(y); subplot(212); plot(y2)
[b,a,k] = LVCas2Dir(Bc,Ac,Gain)
```

1.9.3   LINEAR PHASE FORM

Since linear phase filters have symmetric or anti-symmetric coefficients, delay outputs destined for the same coefficient are combined prior to multiplication. This saves about half the multiplications need for a Direct Form implementation. Figure 1.18 shows a linear phase filter of length-5 having symmetrical coefficients, and Fig. 1.19 shows its equivalent linear phase implementation.

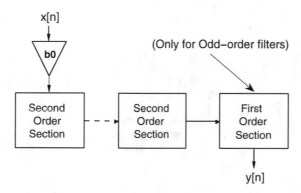

Figure 1.16: A basic cascade arrangement to implement an FIR; each second order section consists of a second order FIR implemented in Direct Form. For odd-order filters, there is one additional first order section.

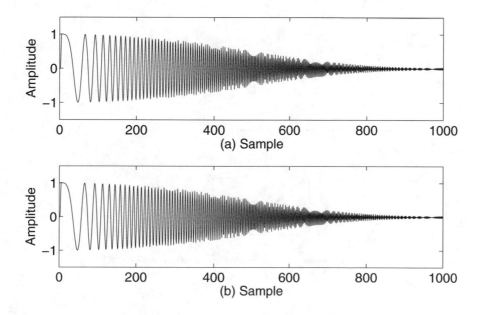

Figure 1.17: (a) A linear chirp filtered using a Direct Form lowpass filter; (b) Same, but filtered using the equivalent Cascade Form filter.

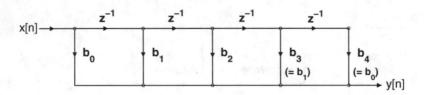

Figure 1.18: A length-5 FIR implemented using Direct Form. Note that the coefficients are symmetrical, having $b_3 = b_1$ and $b_4 = b_0$, so the signals from taps 3 and 4 can be combined with those from taps 0 and 1, respectively, and thus only two multiplications and two additions for the four coefficients b_0, b_1, b_3, and b_4 must be performed rather than the original four multiplications. This more efficient arrangement is shown in Fig. 1.19.

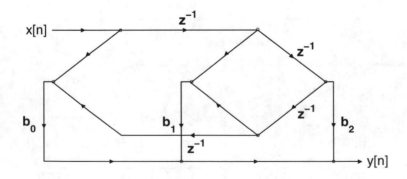

Figure 1.19: A Linear Phase Form filter arrangement for a symmetrical filter having $N = 5$. This form requires only three multiplications instead of the five required in the direct form implementation shown in the previous figure.

1.9.4 CASCADED LINEAR PHASE FORM

Another possibility for a linear phase FIR is to form a cascade of linear phase, real coefficient sections. A typical section would be fourth order, based around a linear phase quad of zeros, although for certain filters, some zeros might come in complex conjugate pairs lying on the unit circle, which would yield second-order real coefficient sections. There can also be single real zeros (of magnitude 1.0) as well as pairs of reciprocal-magnitude real zeros.

1.9.5 FREQUENCY SAMPLING

Recall the formula for reconstructing the z-transform of a sequence $x[n]$ of length N from N samples of the z-transform $\widetilde{X}[k]$ located at frequencies $2\pi k/N$ where $k = 0{:}1{:}N-1$.

$$X(z) = \frac{1 - z^{-N}}{N} \sum_{k=0}^{N-1} \frac{\widetilde{X}[k]}{1 - e^{j2\pi k/N} z^{-1}} \tag{1.8}$$

This form of the z-transform can be used to construct a time domain filter as a cascade of an FIR having the z-transform

$$\frac{1 - z^{-N}}{N}$$

followed by a parallel structure of IIRs of the form

$$\frac{\widetilde{X}[k]}{1 - e^{j2\pi k/N} z^{-1}}$$

The z-transform samples that are not real-only form complex conjugate pairs, and hence they can be collected to make second-order real coefficient sections. The result is that the parallel structure of IIRs uses only real coefficients, which greatly simplifies implementation. A real coefficient filter impulse response $h[n]$ of length N $(0 \le n \le N - 1)$ has a z-transform in the Frequency Sampling Form of

$$H(z) = \frac{1 - z^{-N}}{N} \left[\sum_{k=1}^{M} 2\,|H[k]|\, H_k(z) + \frac{H[0]}{1 - z^{-1}} + \frac{H[N/2]}{1 + z^{-1}} \right] \tag{1.9}$$

where $M = N/2 - 1$ for N even and $M = (N-1)/2$ for N odd, and $H_k(z)$ $(k = 1,1,...M)$ are second order real coefficients sections as follows:

$$H_k(z) = \frac{\cos[\angle H[k]] - \cos[\angle H[k] - (2\pi k/N)]z^{-1}}{1 - 2\cos[2\pi k/N]z^{-1} + z^{-2}} \tag{1.10}$$

In Eq. (1.9), $H[0]$ is real and if N is odd, the term $H[N/2]/(1 + z^{-1})$ will not be present. Note that the values $H[k]$ are z-transform values along the unit circle at DFT frequencies and hence may be computed as the DFT of $h[n]$.

Note that each IIR has a pole of magnitude 1.0, and hence is unstable. To overcome this, the poles and zeros are given magnitude r, slightly less than 1.0, resulting in the following formula:

$$X(z) = \frac{1 - r^N z^{-N}}{N} \sum_{k=0}^{N-1} \frac{\widetilde{X}[k]}{1 - r e^{j2\pi k/N} z^{-1}} \tag{1.11}$$

Example 1.12.   Implement the FIR whose impulse response is $[0.5, 1, 1, 0.5]$ using the Frequency Sampling method.

We need to obtain $H[0]$, $H[1]$, and $H[2]$, which can be done by computing the DFT of the impulse response, which is $[\ 3,\text{-}0.5^*[1\text{+}j],0,\text{-}0.5^*[1\text{-}j]\ ]$.

In general, for a length-4 impulse response, we will have one second order IIR section and two real, first order sections for $H[0]$ and $H[2]$. In this particular case, $H[2] = 0$. To keep things simple, we'll use $r = 1.0$.

The net z-transform will be

$$H(z) = \frac{1 - z^{-4}}{4} \left[\left(\sum_{k=1}^{1} 2(|H[k]|) H_k(z) \right) + \frac{3}{1 - z^{-1}} + \frac{0}{1 + z^{-1}} \right]$$

$$H(z) = \frac{1 - z^{-4}}{4} \left[\left(\sum_{k=1}^{1} 2(0.707) H_k(z) \right) + \frac{3}{1 - z^{-1}} \right]$$

with

$$H_1(z) = \frac{\cos[\angle H[1]] - \cos[\angle H[1] - (2\pi(1)/4)] z^{-1}}{1 - 2\cos[2\pi(1)/4] z^{-1} + z^{-2}}$$

$$H_1(z) = \frac{\cos[5\pi/4] - \cos[5\pi/4 - \pi/2] z^{-1}}{1 - 2\cos[\pi/2] z^{-1} + z^{-2}}$$

which reduces to

$$H_1(z) = \frac{-0.707 + 0.707 z^{-1}}{1 + z^{-2}} = 0.707 \left[\frac{-1 + z^{-1}}{1 + z^{-2}} \right]$$

and the final net z-transform will be

$$H(z) = \frac{1 - z^{-4}}{4} \left[\frac{-1 + z^{-1}}{1 + z^{-2}} + \frac{3}{1 - z^{-1}} \right]$$

Figure 1.20 shows the topology or layout of the filter; note that since $H[2] = 0$, the lowermost IIR would not be implemented in practice. For certain filters that have a large number of $H[k] = 0$, the Frequency Sampling Form implementation can be much more efficient than other implementations.

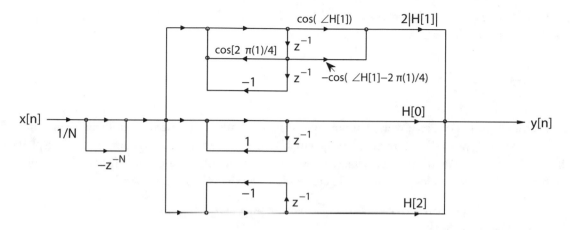

Figure 1.20: The layout of a Frequency Sampling Form equivalent for the simple impulse response [0.5,1,1,0.5]. Note that for this case, $H[2] = 0$, so the lowermost IIR does not need to be implemented.

To verify that Fig. 1.20 is correct, we can process an impulse in the FIR portion

$$\frac{1 - z^{-4}}{4}$$

and then process the result in each of the two nonzero IIRs, summing their outputs. The result should be the original impulse response [0.5,1,1,0.5].

> **unitImp = [1,zeros(1,6)]; y1 = filter([0.25*[1,0,0,0,-1]],[1],unitImp);**
> **y2 = filter([-1,1],[1,0,1],y1); y3 = filter([3],[1,-1],y1);**
> **y = y2 + y3**

References [3] and [4] discuss FIR realizations in detail; [5] gives a very detailed discussion of the Frequency Sampling implementation.

1.10 REFERENCES

[1] T. W. Parks and C. S. Burrus, *Digital Filter Design*, John Wiley & Sons, New York, 1987.

[2] James H. McClellan et al, *Computer-Based Exercises for Signal Processing Using MATLAB 5*, Prentice-Hall, Upper Saddle River, New Jersey, 1998.

[3] John G. Proakis and Dimitris G. Manolakis, *Digital Signal Processing, Principles, Algorithms, and Applications, Third Edition*, Prentice-Hall, Upper Saddle River, New Jersey, 1996.

[4] Vinay K. Ingle and John G. Proakis, *Digital Signal Processing Using MATLAB V.4*, PWS Publishing Company, Boston, 1997.

[5] Richard G. Lyons, *Understanding Digital Signal Processing, Second Edition*, Prentice-Hall, Upper Saddle River, New Jersey 2004.

1.11   EXERCISES

1. Write a script that can receive as an input argument an impulse response that conforms to any one of linear phase Types I-IV, correctly identify which type has been input, and compute and display the amplitude and phase responses as well as the magnitude and phase responses. Your script should conform to the syntax below, and it should create plots like that of Fig. 1.8, for example.

```
function [Type] = LVxAmp_V_MagResponse(Imp,FreqRange,...
IncMag,LogPlot)
% Imp is a linear phase impulse response of Types I, II, III, or IV
% Pass FreqRange as 0 for -pi to pi; 1 for -2pi to 2pi;
% 2 for 0 to pi; 3 for 0 to 2pi, and 4 for 0 to 4pi
% Pass IncMag as 1 to include magnitude plots, or 0 for
% amplitude plots only. Pass LogPlot as 0 for linear
% magnitude plot or 1 for 20log10(Mag) plot;
% The output variable Type is returned as 1,2,3, or 4 for
% Types I, II, III, or IV, respectively, or 0 if Imp is not
% a linear phase impulse response.
% Test calls:
% [Type]= LVxAmp_V_MagResponse([1,0,1],0,1,0)
% [Type]= LVxAmp_V_MagResponse([1,1,1,1],0,1,0)
% [Type]= LVxAmp_V_MagResponse([1,0,-1],0,1,0)
% [Type]= LVxAmp_V_MagResponse([1,1,-1,-1],0,1,0)
```

2. Write a script that conforms to the following call syntax:

```
function LVxMovingAverageFilter(MALength,NoiseAmp)
% MALength is the length of the Moving Average filter
% NoiseAmp is the StdDev of white noise mixed with a coherent
% signal, which consists of five rectangular pulses over 1024
% samples, each coherent pulse having an amplitude of
% 1.0 and a width of 20 samples.
% Test calls:
% LVxMovingAverageFilter(5,0.8)
% LVxMovingAverageFilter(10,0.8)
% LVxMovingAverageFilter(20,0.8)
% LVxMovingAverageFilter(40,0.8)
```

3. A signal consists of a one millisecond long positive pulse followed by a one millisecond long negative pulse, immersed in random noise. The signal repeats itself every 20 milliseconds. Your receiver samples at a rate of 10 kHz, and the total signal duration is one second.

(a) Devise a suitable matched FIR filter to enhance the signal.

(b) Design a recursive algorithm to implement the filter designed in (a).

(c) Write a script that generates the test signal with a user-specified amount of noise, and filters the test signal using (1) the matched filter of (a) above using function *conv*, and (2) the recursive algorithm of (b) above. While the ideal matched filter will precisely match the sought-after signal in duration as well as shape, it is instructive to experiment with filters of the proper shape but differing durations. Thus, the script should allow, using the input argument *lenMDfilt*, the number of samples in the filter to be readily changed. After filtering the test signal using both methods, plot the test signal and both filtering results for various levels of noise and various filter lengths as specified in the test calls given below in the following function specification. Figure 1.21 is an example of the result for the call

LVxMDfilter(0.75,20,2000)

```
function LVxMDfilter(k,lenMDfilt,plotlim)
% Creates a test signal sampled at 10 kHz consisting of, over
% a duration of one second, fifty equally spaced bipolar pulses
% each consisting of a one millisecond long positive pulse
% followed immediately by a one millisecond long negative
% pulse, the entire one second test signal consisting of the fifty
% bipolar pulses plus white noise having standard deviation
% equal to k. A matched FIR filter of length lenMDfilt is used
% to improve the signal-to-noise ratio, i.e., to emphasize the
% signal relative to the noise. plotlim is the number of samples
% of tstSig and the two filtered sequences to plot
% Test calls:
% LVxMDfilter(0.1,20,2000)
% LVxMDfilter(0.25,20,2000)
% LVxMDfilter(0.5,20,2000)
% LVxMDfilter(1,20,2000)
% LVxMDfilter(2,20,2000)
% LVxMDfilter(0.1,6,2000)
% LVxMDfilter(0.25,6,2000)
% LVxMDfilter(0.5,6,2000)
% LVxMDfilter(1,6,2000)
% LVxMDfilter(2,6,2000)
```

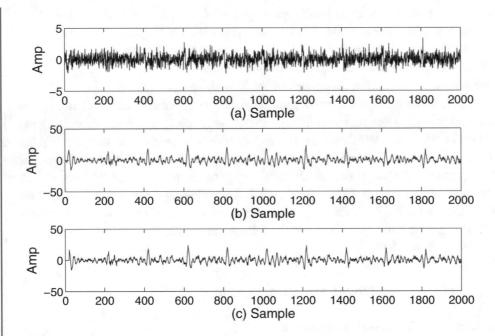

Figure 1.21: (a) Test signal, bipolar pulses in white noise (see text); (b) Test signal filtered with a matched filter using the function *conv*; (c) Test signal filtered with the matched filter using a recursive algorithm.

4. For each of the sets of conditions listed below, answer the following questions using the VI *DemoDragZeroZxformVI*:

Questions
i) Describe the resultant impulse response as real or complex
ii) Describe the phase response as linear or nonlinear
iii) Note the minimum and maximum values of the magnitude of the z-transform and the corresponding frequencies at which these occur
iv) Characterize the passband type of the filter i.e., lowpass, highpass, etc.
v) Symmetry of magnitude response about frequency zero

Conditions:
a) *Mode Select = Single Zero* and move the zero cursor Z to approximately ($\approx$) $1 + 0j$
b) *Mode Select = Complex Conjugate Zeros*; $Z \approx 0 + j$
c) *Mode Select = Complex Conjugate Zeros*; $Z \approx 0 - j$
d) *Mode Select = Single Zero*; $Z \approx -1$
e) *Mode Select = Complex Conjugate Zeros*; $Z \approx 0 + j$
f) *Mode Select = Complex Conjugate Zeros*; $Z \approx 0.9 + 0.1j$
g) *Mode Select = Quad Zeros*; $Z \approx 0.9 + 0.1j$

h) *Mode Select = Complex Conjugate Zeros; Z ≈ 0 + 0.9 j*
i) *Mode Select = Quad Zeros; Z ≈ 0 + 0.9 j*
j) *Mode Select = Complex Conjugate Zeros; Z ≈ -0.9 + 0.1 j*
k) *Mode Select = Quad Zeros; Z ≈ -0.9 + 0.1 j*
l) *Mode Select = Quad Zeros; Z ≈ 0.6 + 0.6 j*
m) *Mode Select = Complex Conjugate Zeros; Z ≈ -0.6 + 0.6 j*
n) *Mode Select = Quad Zeros; Z ≈ 0.6 + 0.6 j*
o) *Mode Select = Complex Conjugate Zeros; Z ≈ -0.6 + 0.6 j*
p) *Mode Select = Single Zero; Z ≈ 0.707 + 0.707 j*
q) *Mode Select = Single Zero; Z ≈ 0.707 - 0.707 j*

5. Design the shortest linear phase FIR that has [0.65 + 0.65j] as one of its zeros. Evaluate the magnitude and phase response of the resultant linear phase filter to verify its phase linearity.

6. (a) Design a comb filter having impulse response = [1,zeros(1, N),1] that is to give the maximum attenuation possible to a 60 Hz cosine wave sampled at 44,100 Hz. Determine N, then plot the log-magnitude spectrum of the signal after being filtered by your comb filter. Repeat the filtering and log-magnitude plot using values of N from three less to three more than the value you computed to verify that your value of N gives the best attenuation.

(b) Repeat part (a), this time using as the signal to be attenuated the following:

$$y = \cos(120\pi t) + \cos(360\pi t) + \cos(600\pi\ t)$$

where $t = [0:1/(44,099):1]$. Compare results to those of the original problem using only the 60 Hz signal.

(c) Repeat part (a), but use a subtractive comb filter of the following form: [1, zeros(1,M),-1]. You should find that the ultimate attenuation achievable is higher using this comb filter rather than that of part (a) above. Why?

7. Compute and plot the amplitude and magnitude responses of the following linear phase filters, represented by their impulse responses:

(a) [1,zeros(1,7),1]
(b) [1,zeros(1,6),1]
(c) [1,1,zeros(1,5),-1,-1]
(d) [1,1,zeros(1,4),-1,-1]
(e) [-0.0052,-0.0229,0.0968,0.4313,0.4313,0.0968,-0.0229,-0.0052]

8. Compute and plot the system (or z-transform) zeros for each of the impulse responses in the previous problem, and identify single real zeros, complex conjugate zero-pairs, and zero-quads.

9. Verify that the amplitude response for a Type-III linear phase filter, given by

$$H_r(\omega) = 2 \sum_{n=0}^{M-1} h[n] \sin(\omega[M - n]) \tag{1.12}$$

(where $M = (L - 1)/2$ and L is the filter length) is correct for the impulse response

$$Imp = [a, b, c, 0, -c, -b, -a]$$

by obtaining an expression for the DTFT of Imp and modifying the expression until it is in the form given by Eq. (1.12).

10. Verify that the amplitude response for a Type-IV linear phase filter, given by

$$H_r(\omega) = 2 \sum_{n=0}^{L/2-1} h[n] \sin(\omega[M - n])$$

is correct for the impulse response

$$Imp = [a, b, c, -c, -b, -a]$$

11. Write a script to implement Eqs. (1.9) and (1.10), i.e., to convert a set of Direct Form FIR coefficients into a set of coefficients for a Frequency Sampling implementation, according to the following specification:

> **function [CFsB,CFsA,BFs,AFs] = LVxDirect2FreqSampFIR(Imp)**
> **% Receives an FIR impulse response Imp and generates**
> **% the Frequency Sampling Coefficients BFs,and AFs with**
> **% the comb filter section coefficients as CFsB and CFsA.**
> **% Test calls:**
> **% [CFsB,CFsA,BFs,AFs] = LVxDirect2FreqSampFIR([1,1,1,1])**
> **% [CFsB,CFsA,BFs,AFs] = LVxDirect2FreqSampFIR([1,-1,1,-1])**
> **% [CFsB,CFsA,BFs,AFs] = LVxDirect2FreqSampFIR([1,0,0,1])**

12. Write a script that receives a set of Frequency Sampling Form coefficients corresponding to an impulse response Imp and filters a signal x using both the Direct Form coefficients (i.e., Imp itself) and the Frequency Sampling Form coefficients. Display the results of both filtering operations. Follow the function specification below:

> **LVxFreqSampFilter(Imp,CFsB,CFsA,BFs,AFs,x)**
> **% Receives an impulse response Imp and a signal x, filters x**
> **% and displays the result two different ways, first, using the**
> **% impulse response itself as a Direct Form FIR, and second,**
> **% using Frequency Sampling implementation coefficients**
> **% CFsB,CFsA,BFs,AFs. (created for example, by the script**
> **% LVxDirect2FreqSampFIR).**

The m-code

Imp = [1,1,1,1]; x = chirp([0:1/999:1],0,1,500);
[CFsB,CFsA,BFs,AFs] = LVxDirect2FreqSampFIR(Imp)

LVxFreqSampFilter(Imp,CFsB,CFsA,BFs,AFs,x)

should, for example, result in Fig. 1.22.

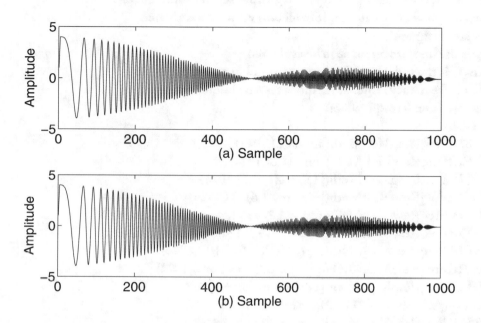

Figure 1.22: (a) Convolution of a linear chirp of length 1000 with the Direct Form coefficients [1,1,1,1]; (b) Result from filtering the linear chirp with the Frequency Sampling implementation of the Direct Form coefficients [1,1,1,1].

Additional test m-code:

```
Imp = fir1(22,0.3); x = chirp([0:1/999:1],0,1,500);
[CFsB,CFsA,BFs,AFs] = LVxDirect2FreqSampFIR(Imp)
LVxFreqSampFilter(Imp,CFsB,CFsA,BFs,AFs,x)

Imp = fir1(22,[0.3,0.5]); x = chirp([0:1/999:1],0,1,500);
[CFsB,CFsA,BFs,AFs] = LVxDirect2FreqSampFIR(Imp)
LVxFreqSampFilter(Imp,CFsB,CFsA,BFs,AFs,x)

Imp = fir1(82,[0.4,0.6],'stop'); x = chirp([0:1/999:1],0,1,500);
[CFsB,CFsA,BFs,AFs] = LVxDirect2FreqSampFIR(Imp)
LVxFreqSampFilter(Imp,CFsB,CFsA,BFs,AFs,x)
```

13. Write a script that will evaluate the frequency response of an FIR from 0 to 2π radians using the DTFT, the z-transform, a real chirp, and a complex chirp, and plot the results, in accordance with the following function specification.

function LVxFIRFreqRespMultMeth(imp,lenEval)

```
% Receives a real or complex impulse response and computes
% and displays the frequency response using four methods, namely
% the DTFT, the z-transform, real chirp response, and complex
% chirp response. The magnitude of all four responses is plotted
% on a single figure.
% imp is the impulse response to be evaluated
% lenEval is the number of frequency samples to compute
% All four frequency response methods test the frequency
% response from 0 to 2pi radians.
% Test calls:
% LVxFIRFreqRespMultMeth(ones(1,8),1024)
% LVxFIRFreqRespMultMeth(-ones(1,8),1024)
% LVxFIRFreqRespMultMeth([1,zeros(1,6),1],1024)
% LVxFIRFreqRespMultMeth([-1,zeros(1,6),-1],1024)
% LVxFIRFreqRespMultMeth( (ones(1,8) + j*[1,zeros(1,6),1]),1024)
% LVxFIRFreqRespMultMeth((ones(1,8) - j*[1,zeros(1,6),1]),1024)
% LVxFIRFreqRespMultMeth(exp(j*2*pi*[0:1:7]*3/8),1024)
% LVxFIRFreqRespMultMeth(exp(-j*2*pi*[0:1:7]*3/8),1024)
% LVxFIRFreqRespMultMeth([exp(-j*2*pi*[0:1:31]*5/32)+...
%   0.5*exp(j*2*pi*[0:1:31]*11/32)],1024)
% LVxFIRFreqRespMultMeth([cos(2*pi*[0:1:31]*5/32)+...
%   0.5*sin(2*pi*[0:1:31]*11/32)],1024)
```

CHAPTER 2

FIR Design Techniques

2.1 OVERVIEW

In the previous chapter we examined a number of general ideas or principles related to FIR design, such as the effect of filter length, the effect of windowing, requirements for linear phase, etc. Additionally, we have gained knowledge of simple filters such as the Comb and Moving Average filters, as well as simple passband filters having arbitrary band limits built by superposing two or more frequency-contiguous correlators (covered in Volume I of the series, see the Preface of this volume for information on the contents of the other volumes in the series).

We have at last accumulated enough knowledge to successfully undertake the design of linear phase FIRs that can meet certain user-specified filter design requirements, including particular passband and stopband boundaries, particular levels of stopband attenuation, passband ripple, etc. After presenting a brief overview of the three main design methods that will be explored in this chapter, we set forth the various standard parameters used to specify desired filter characteristics, which will allow us to not only specify the requirements for a given filter, but to evaluate and compare the performance of different filters designed to meet the same criteria. We then launch into a detailed discussion of the windowed ideal lowpass filter technique, including how to generate highpass, bandpass, and bandstop filters from lowpass filters. This is followed by an examination of the Frequency Sampling Design Method (not to be confused with the Frequency Sampling Realization Method discussed in the previous chapter with regard to the realization, rather than the design, of FIRs), which uses the inverse DFT to generate an impulse response from a user-specified set of frequency domain samples. With this design method we'll also explore the use of optimized transition band sample amplitudes or coefficients, a simple technique that can greatly improve stopband attenuation. We'll also investigate the design of certain linear phase Type III and Type IV specialty filters, the Hilbert transformer and the differentiator. The last major topic in the chapter is a very important one, equiripple FIR design; the equiripple filter design technique, although somewhat difficult to understand and implement compared to the windowed-lowpass and frequency sampling design techniques, is very popular since it results in filters that can achieve a given design with the shortest length.

2.2 SOFTWARE FOR USE WITH THIS BOOK

The software files needed for use with this book (consisting of m-code (.m) files, VI files (.vi), and related support files) are available for download from the following website:

http://www.morganclaypool.com/page/isen

The entire software package should be stored in a single folder on the user's computer, and the full file name of the folder must be placed on the MATLAB or LabVIEW search path in accordance with the instructions provided by the respective software vendor (in case you have encountered this notice before, which is repeated for convenience in each chapter of the book, the software download only needs to be done once, as files for the entire series of four volumes are all contained in the one downloadable folder). See Appendix A for more information.

2.3   SUMMARY OF DESIGN METHODS

Three standard methods to design an FIR which will be covered in this chapter are:

- **The Window Method**: In this method, a truncated ideal lowpass filter having a certain bandwidth is generated, and then a chosen window is applied to achieve a certain stopband attenuation. Filter length can be adjusted to achieve a needed roll-off rate in the transition band. Filter types other than lowpass, such as highpass, bandpass, and notch, can be achieved by several techniques that start with windowed, truncated ideal lowpass filters.

- **The Frequency Sampling Method**: In this method, evenly-spaced samples of a desired frequency response are created, and the IDFT is computed to obtain an impulse response. In addition to using the IDFT to accomplish this, there are formulas that do the equivalent, i.e., superposing orthogonal cosines or sines to generate an impulse response. Both variants of this method will be explored below. Rather than creating an impulse response to be implemented either in Direct Form or Linear Phase Form, it is possible to use the frequency samples to directly realize the filter using this method's namesake, the Frequency Sampling Realization Method, which will be explored in the exercises at the end of the chapter.

- **The Equiripple Method**: This method designs an FIR having equalized ripple amplitudes in the passband, and equalized ripple in the stopband. The ripple levels may be independently controlled, allowing great flexibility. It is possible with equiripple design, for example, for a given filter length, to increase stopband attenuation by letting passband ripple increase, and vice versa. The previous two FIR design methods do not permit this degree of control.

2.4   FILTER SPECIFICATION

Figure 2.1 illustrates a typical design specification for an FIR lowpass filter. There is generally a certain amount of ripple in both passbands and stopbands. In Fig. 2.1, the maximum deviation (considered as an acceptable tolerance) from the average value in the passband is designated δ_P, while the deviation from zero in the stopband is designated as δ_S. This manner of defining the requirements for passband and stopband ripple is called an absolute specification; another manner (more common) is to specify the levels of ripple in decibels relative to the maximum magnitude of response, which is $1 + \delta_P$.

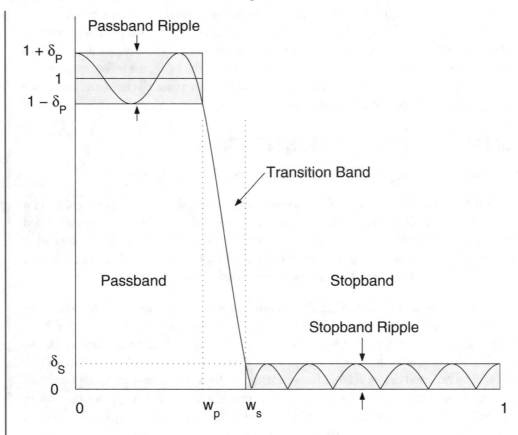

Figure 2.1: Design criteria for an FIR.

Figure 2.2 depicts another lowpass filter design specification in relative terms. The values of R and A are in decibels, and represent the passband ripple amplitude (or minimum passband response when the maximum filter/passband response is 0 db) and minimum stopband attenuation, respectively. The relationship between δ_P and δ_S and R and A are

$$R = -20 \log 10(\frac{1 - \delta_P}{1 + \delta_P})$$

and

$$A = -20 \log 10(\frac{\delta_S}{1 + \delta_P})$$

To determine δ_P when R and A are given, use

$$\delta_P = \frac{1 - 10^{-R/20}}{1 + 10^{-R/20}}$$

and

$$\delta_S = (1 + \delta_P)10^{-A/20}$$

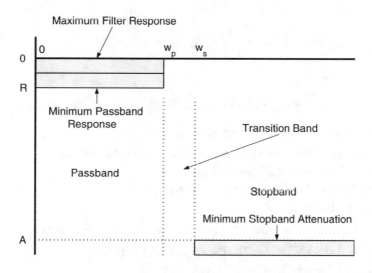

Figure 2.2: A relative filter design specification, with the (horizontal) frequency axis at the top, and (vertical) logarithmic amplitude (dB) axis at the left.

Example 2.1.   A certain filter design specification is expressed in absolute terms as $\delta_P = 0.02$ and $\delta_S = 0.002$. Determine the filter design specification in relative terms.

We have

$$R = -20 \log 10(\frac{0.98}{1.02}) = 0.3475 \; db$$

and

$$A = -20 \log 10(\frac{0.002}{1.02}) = 54.15 \; db$$

In terms of m-code, we get

R = -20*log10(0.98/1.02)
A = -20*log10(0.002/1.02)

Example 2.2.   A certain filter design specification is expressed in relative terms as $R = 0.5$ db and $A = 60$ db. Determine the absolute specification.

We get

$$\delta_P = \frac{1 - 10^{-0.5/20}}{1 + 10^{-0.5/20}} = 0.028774$$

and

$$\delta_S = (1.028774)(10^{-60/20}) = 0.001028$$

Suitable m-code to compute δ_P and δ_S is

function [DeltaP,DeltaS] = LVRelSpec2AbSpec(Rp,As)
% [DeltaP,DeltaS] = LVRelSpec2AbSpec(0.5,60)
Rfac= 10^(-Rp/20); DeltaP = (1-Rfac)/(1+Rfac);
DeltaS = (1+DeltaP)*10^(-As/20);

To check the computation, use this m-code to return to relative specification:

function [Rp,As] = LVAbSpec2RelSpec(DeltaP,DeltaS)
% [Rp,As] = LVAbSpec2RelSpec(DeltaP,DeltaS)
Rp = -20*log10((1-DeltaP)/(1+DeltaP));
As = -20*log10(DeltaS/(1+DeltaP));

2.5   FIR DESIGN VIA WINDOWED IDEAL LOWPASS FILTER

An Ideal Lowpass filter has a noncausal, infinite-length impulse response which can be determined by taking the Inverse DTFT of the frequency specification

$$X(e^{j\omega}) = \begin{cases} 1 \cdot e^{-j\omega M} & |\omega| \leq \omega_c \\ 0 & |\omega| > \omega_c \end{cases}$$

which can be written as

$$x[n] = \frac{1}{2\pi} \int_{-\omega_c}^{\omega_c} e^{-j\omega M} e^{jwn} d\omega = \frac{1}{2\pi} \int_{-\omega_c}^{\omega_c} e^{j\omega(n-M)} d\omega = \frac{\sin(\omega_c[n-M])}{\pi[n-M]} \tag{2.1}$$

To utilize such an impulse response for a linear phase FIR, it is necessary to symmetrically truncate it about M, where $M = (L-1)/2$ and L is the total length of the symmetrically-truncated impulse response. Since the impulse response is symmetrical, Type I and II linear phase filters can be designed.

Example 2.3.   Write m-code that will generate a causal, symmetrically truncated impulse response of the ideal lowpass type; compute and plot for $\omega_c = 0.25$ and $L = 61$.

The following m-code generates such an impulse response; change w_c to control cutoff frequency and L to control impulse response length. Note that a true ideal lowpass filter impulse response is a continuous function of infinite duration from $t = -\infty$ to $+\infty$; in this example, we are computing a finite number of samples of such a function. For the purposes of bandlimited discrete processing and digital filtering, of course, this is acceptable.

```
function b = LVIdealLPFImpResp(wc,L)
% LVIdealLPFImpResp(0.25*pi,61)
M = (L-1)/2; n = 0:1:L-1;
b = sin(wc*(n - M + eps))./(pi*(n - M + eps));
figure(55); stem(n,b)
```

2.5.1   WINDOWS

Any finite length (i.e., truncated) version of the ideal lowpass impulse response may be considered as the product of the infinite-length lowpass impulse response and a window function W, which has a finite number of contiguous nonzero-valued samples

$$b = (\frac{\sin(\omega_c[n - M])}{\pi[n - M]})(W_L[n - M]) \tag{2.2}$$

where the window length is L, $M = (L - 1)/2$, $0 \leq n \leq L - 1$, and $W_L[n]$ is generally a function $F_E[n]$ having even symmetry about M defined as

$$W_L[n] = \begin{cases} F_E[n] & n = 0:1:L - 1 \\ 0 & \text{otherwise} \end{cases}$$

The right side of Eq. (2.2) is the product of an infinite-length, ideal lowpass filter with a function that is nonzero only over a finite number of contiguous samples. The result is a finite-length or truncated lowpass filter. Since W is chosen to have even symmetry around M, the window symmetrically truncates the infinite-length function (which is itself symmetrical about M), resulting in a finite-length symmetrical sequence as the net filter impulse response.

We encountered windows in Volume II of the series as a tool for reducing leakage in the DFT. We discuss some of the basic information on standard windows here for convenience.

The simplest window is the rectangular window $R[n]$, which is defined as

$$R[n] = \begin{cases} 1 & 0 \leq n \leq L - 1 \\ 0 & \text{otherwise} \end{cases}$$

The windowing process, using a rectangular window, is shown in Fig. 2.3. In subplot (a), a portion of an infinite-length, ideal lowpass filter is shown, centered on M; in subplot (b), a portion of the infinite-length (rectangular) window is shown, with its contiguous, nonzero-valued samples centered on M, and finally, in subplot (c), the product of the two sequences in (a) and (b), the truncated lowpass filter, is shown.

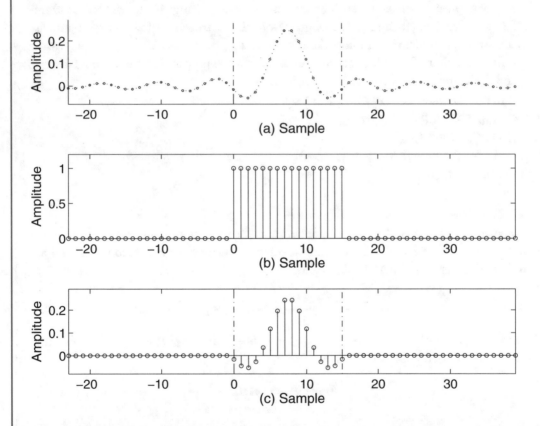

Figure 2.3: (a) Samples of an Ideal Lowpass filter impulse response, delayed by $M = (L - 1)/2$, where L is the desired digital filter length; (b) A Rectangular window of length L; (c) The product of the window and the Ideal Lowpass filter, yielding, for samples $n = 0{:}1{:}(L - 1)$, a length-L symmetrical impulse response as the lowpass digital filter finite impulse response.

The Hanning, Hamming, and Blackman windows are described by the general raised-cosine formula

$$W[n] = \begin{cases} a - b\cos(2\pi \frac{n}{L-1}) + c\cos(4\pi \frac{n}{L-1}) & n = 0:1:L-1 \\ 0 & \text{otherwise} \end{cases}$$

where L is the window length, and where, for the Hanning window, $a = 0.5$, $b = 0.5$, and $c = 0$; for the Hamming window, $a = 0.54$, $b = 0.46$, and $c = 0$; and for the Blackman window, $a = 0.42$, $b = 0.5$, and $c = 0.08$.

Figure 2.4 shows the windowing process using a Hamming window:

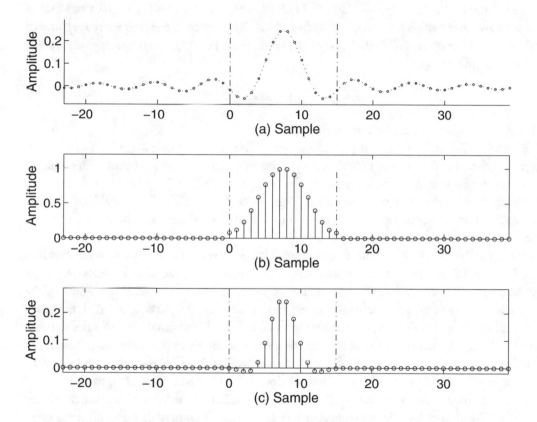

Figure 2.4: (a) Samples of an Ideal Lowpass filter impulse response, delayed by $M = (L - 1)/2$, where L is the desired digital filter length; (b) A Hamming window of length L; (c) The product of the window and the Ideal Lowpass filter, yielding, for samples $n = 0:1:(L - 1)$, a length-L symmetrical impulse response as the lowpass digital filter finite impulse response.

The Kaiser window is described by the formula

$$W[n] = \begin{cases} \frac{I_0[\beta(1-(n-M)/M]^2)^{0.5}]}{I_0(\beta)} & n = 0 : 1 : L - 1 \\ 0 & \text{otherwise} \end{cases}$$

for $n = 0:1:L-1$, where L is the window length, $M = (L - 1)/2$, and I_0 represents the modified Bessel function of the first kind.

2.5.2   NET FREQUENCY RESPONSE

The net frequency response of the impulse response b is the circular convolution of the DTFTs of the ideal lowpass filter (an infinite-length time domain sequence) and the window (of infinite length but containing nonzero values only over the interval $0 \leq n \leq L - 1$). This process results in a smearing or widening of the frequency response (i.e., the DTFT) of the net lowpass filter relative to that of the ideal lowpass filter, which has infinitely steep roll-off. Representing the frequency response of the ideal lowpass filter by $H_L(e^{j\omega})$, and that of the window by $W(e^{j\omega})$, the frequency response of the windowed ideal lowpass filter is

$$H(e^{j\omega}) = H_L(e^{j\omega}) \circledast W(e^{j\omega})$$

where the symbol $\circledast$ here means circular convolution.

Figures 2.5, 2.6, 2.7, and 2.8, depict this process. The generic process is shown in Fig. 2.5. The frequency domain effect of the time domain process of windowing (whereby an infinite length ideal lowpass filter impulse response is symmetrically truncated to a finite length) can be determined by numerically performing circular convolution of samples of the DTFTs of the ideal lowpass filter and the proposed window; the ripple and transition width of the window determine the ultimate frequency response of the truncated lowpass filter.

Looking at the computational process in more detail for several different windows, for each of Figs. 2.6, 2.7, and 2.8, the computation started with a good approximation of an ideal lowpass filter (an impulse response having a length of thousands of samples) and a symmetrical window of the same length consisting of the value zero everywhere except in the central portion, in which is located a symmetrical group of contiguous, nonzero samples of a desired length which form the window (Hamming, Kaiser, etc) which will truncate the ideal lowpass filter's impulse response. The DFT (i.e., samples of the DTFT) is obtained of the ideal lowpass filter impulse response (approximated by a very large number of samples) and of the window, which is of the same length as the ideal lowpass filter, and then the circular convolution of the two DFTs is obtained, which is the frequency response of the ideal lowpass filter as truncated by the window. The magnitude of all three (very long) DFTs is then plotted.

- Each type of window, when applied to a truncated ideal lowpass impulse response, results in a filter having a characteristic main lobe width, transition width, and minimum stopband attenuation, the first two of which depend strongly on filter length.

- The minimum stopband attenuation is generally quoted as a constant value. As a result, it is possible to choose a window according to what minimum level of stopband attenuation is acceptable, and then adjust L (filter length) to achieve the needed roll-off or narrowness of transition band.

- The rectangular window has the narrowest main lobe and smallest transition width, but the poorest stopband attenuation.

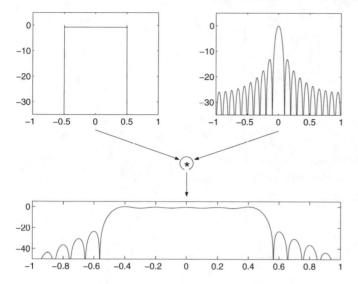

Figure 2.5: Upper left plot: samples of the DTFT (obtained via DFT) of an ideal lowpass filter having $\omega_c = 0.5\pi$ radians, showing to good approximation a very steep roll-off (theoretically infinitely steep); Upper right plot: samples of the DTFT of a rectangular window, the rectangular window having been used to symmetrically truncate the ideal lowpass filter's impulse response; Lower plot: the net frequency response of the truncated lowpass filter, computed as the circular convolution of the DTFTs of the ideal lowpass filter and the rectangular window. All horizontal axes are frequency in units of π, and all vertical axes are magnitude in dB.

- The Hanning, Hamming, and Blackman windows have broader main lobes, wider transition widths, but improved stopband attenuation.

- The Kaiser window is adjustable, allowing a chosen compromise between transition width and stopband attenuation through the choice of the parameter β.

- Approximate and exact values of the transition width as a function of window length for the standard windows have been tabulated and can be used to give a good first estimate of the needed filter length. The following table gives approximate and exact values for L in terms of the transition width, $\omega_t = w_s - w_p$ and the minimum stopband attenuation values for several standard windows. The exact values for L are better estimates, and can be used when the design target A_s is close to the window's inherent A_s. If the target A_s is much smaller than the window's inherent A_s, the filter length needed may be shorter.

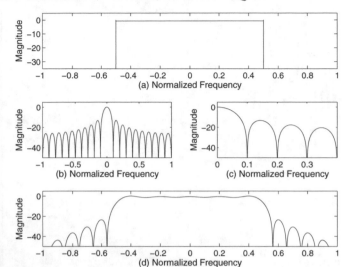

Figure 2.6: (a) Magnitude (dB) of DFT of an ideal lowpass filter ($L = 2^{15}$ samples); (b) Magnitude (dB) of DFT of a window of length 2^{15} samples, all which are zero except for the central 20 samples, which form a rectangular window; (c) Zoomed-in view of waveform at (b); (d) Net frequency response (magnitude in dB) of the truncated ideal lowpass filter, computed as the circular convolution of the DFTs of the ideal lowpass filter and the window.

Name	Approx L	Exact L	min A_s, dB
Blackman	$12\pi/\omega_t$	$11\pi/\omega_t$	74
Hamming	$8\pi/\omega_t$	$6.6\pi/\omega_t$	53
Hanning	$8\pi/\omega_t$	$6.2\pi/\omega_t$	44
Bartlett	$8\pi/\omega_t$	$6.1\pi/\omega_t$	25
Rectangular	$4\pi/\omega_t$	$1.8\pi/\omega_t$	21

The Kaiser window is adjustable according to the parameter β, and Kaiser has provided empirical formulas that allow determination of necessary values of L (filter length) and β to achieve a certain minimum stopband attenuation A_s. The needed length L for a given A_s is

$$L \simeq \frac{2\pi(A_s - 7.95)}{14.36(\omega_s - \omega_p)} + 1 \tag{2.3}$$

where ω_s and ω_p are in radians, such as 0.5π, etc., and the needed β is

$$\beta = \begin{cases} 0.1102(A_s - 8.7) & A_s \geq 50 \\ 0.5842(A_s - 21)^{0.4} + 0.07886(A_s - 21) & 21 \leq A_s < 50 \\ 0 & A_s < 21 \end{cases} \tag{2.4}$$

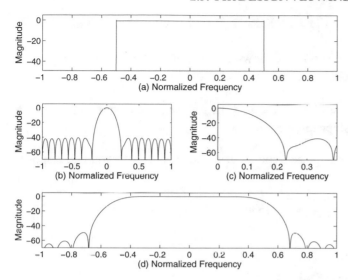

Figure 2.7: (a) Magnitude (dB) of DFT of an ideal lowpass filter ($L = 2^{15}$ samples); (b) Magnitude (dB) of DFT of a window of length 2^{15} samples, all which are zero except for the central 20 samples, which form a Hamming window; (c) Zoomed-in view of waveform at (b); (d) Net frequency response (magnitude in dB) of the truncated ideal lowpass filter, computed as the circular convolution of the DFTs of the ideal lowpass filter and the window.

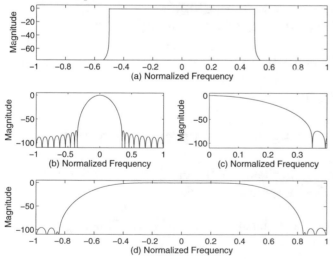

Figure 2.8: (a) Magnitude (dB) of DFT of an ideal lowpass filter ($L = 2^{15}$ samples); (b) Magnitude (dB) of DFT of a window of length 2^{15} samples, all which are zero except for the central 20 samples, which form a Kaiser(10) window; (c) Zoomed-in view of waveform at (b); (d) Net frequency response (magnitude in dB) of the truncated ideal lowpass filter, computed as the circular convolution of the DFTs of the ideal lowpass filter and the window.

2.5.3   WINDOWED LOWPASS FILTERS-PASSBAND RIPPLE AND STOP-BAND ATTENUATION

We illustrate the practical application of various windows, and the effect on passband ripple and stopband attenuation to a lowpass impulse response with several examples:

Example 2.4.   Design a length-17 lowpass filter having $\omega_p = 0.4\pi$ and $\omega_s = 0.5\pi$ using a rectangular window. Measure passband ripple and stopband attenuation.

The function below constructs the impulse response using an ideal lowpass impulse response with a rectangular window. The result from making the call

$$\text{LVLPFViaSincRectwin(0.4*pi,0.5*pi,17)}$$

is shown in Fig. 2.9.

```
function LVLPFViaSincRectwin(wp,ws,L)
% LVLPFViaSincRectwin(0.4*pi,0.5*pi,17)
wc = (wp + ws)/2; M = (L-1)/2; n = 0:1:L-1;
b = sin(wc*(n - M + eps))./(pi*(n - M + eps));
LenFFT = 8192; fr = abs(fft(b,LenFFT)); fr=fr(1,1:LenFFT/2+1);
Lfr = length(fr); PB = fr(1,1:round((wp/pi)*Lfr));
SB = fr(1,round((ws/pi)*Lfr):Lfr);
PBR = -20*log10(min(PB)), SBAtten = -20*log10(max(SB)),
figure(59); plot([0:1:LenFFT/2]/(LenFFT/2), 20*log10(fr+eps));
xlabel('Frequency, Units of \pi');
text(0.1,-30,['actual Rp = ',num2str(PBR,3),' dB'])
text(0.1,-45,['actual As = ',num2str(SBAtten,3),' dB'])
ylabel(['Mag, dB']); axis([0 1 -inf inf])
```

2.5.4   HIGHPASS, BANDPASS, AND BANDSTOP FILTERS FROM LOWPASS FILTERS

To create filters other than lowpass from a lowpass filter, several examples are presented that illustrate the general procedure, which can be described as spectral subtraction.

Example 2.5.   Design a highpass filter using a rectangular window of length 51 having $\omega_c = 0.3\pi$ using a truncated ideal lowpass filter.

To do this, we will design a lowpass filter having the same cutoff (0.3π), and subtract it from a filter of the same length (51) that passes all frequencies from 0 to π radians. The following code illustrates this procedure; the result from making the call

$$\text{LVHPFViaSincLPFRectwin(0.3*pi,51)}$$

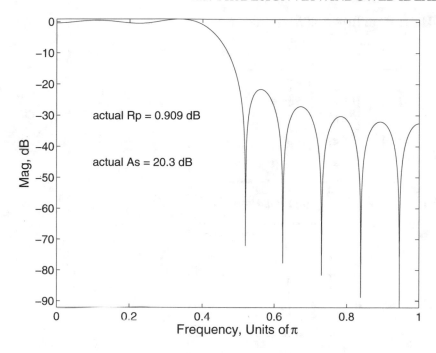

Figure 2.9: The magnitude of the frequency response of a length-17 lowpass filter having an inherent rectangular window.

is shown in Fig. 2.10.

```
function LVHPFViaSincLPFRectwin(wc,L)
% LVHPFViaSincLPFRectwin(0.3*pi,51)
M = (L-1)/2; n = 0:1:L-1;
ImpLo = sin(wc*(n - M + eps))./(pi*(n - M + eps));
ap = sin(pi*(n - M + eps))./(pi*(n - M + eps));
ImpHi = ap - ImpLo; frImpHi = abs(fft(ImpHi,1024));
frImpLo = abs(fft(ImpLo,1024)); figure(56);
set(56,'color',[1,1,1]); subplot(221); stem(n,ImpLo);
xlabel('(a) Sample'); ylabel('Amplitude')
axis([0 length(ImpLo) 1.2*min(ImpLo) 1.1*max(ImpLo)])
subplot(222); plot([0:1:512]/512, frImpLo(1,1:513));
xlabel(['(b) Freq, Units of \pi']); ylabel('Magnitude')
axis([0 1 0 1.1]); subplot(223); stem(n,ImpHi);
xlabel('(c) Sample'); ylabel('Amplitude')
axis([0 length(ImpHi) 1.2*min(ImpHi) 1.1*max(ImpHi)])
subplot(224); plot([0:1:512]/512,frImpHi(1,1:513));
```

xlabel(['(d) Freq, Units of \pi']); ylabel('Magnitude')
axis([0 1 0 1.1])

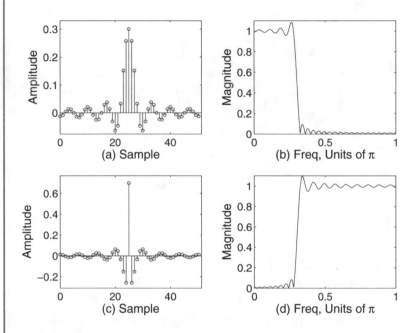

Figure 2.10: (a) Lowpass filter impulse response; (b) Magnitude of DTFT of sequence in (a); (c) Highpass impulse response; (d) Magnitude of DTFT of sequence in (c).

Example 2.6.   Design a bandpass filter of length 79, using a rectangular window, having band edges [0.2, 0.3, 0.5, 0.6], where the first stopband runs from 0 to 0.2π radians, the first transition band runs from 0.2π to 0.3π radians, the second transition band runs from 0.5π to 0.6π radians, and the second stopband runs from 0.6π to π radians.

To do this, we will design a lowpass filter of length 79 having transition band from 0.5π to 0.6π ($\omega_{c2} = (0.5 + 0.6)\pi/2$), then subtract from its impulse response that of a lowpass filter of length 79 having its transition band from 0.2π to 0.3π ($\omega_{c1} = (0.2 + 0.3)\pi/2$). The following code implements this procedure; the result from making the call

LVBPFViaSincLPFRectwin(0.25*pi,0.55*pi,79)

is shown in Fig. 2.11.

```
function LVBPFViaSincLPFRectwin(wc1,wc2,L)
% LVBPFViaSincLPFRectwin(0.25*pi,0.55*pi,79)
M = (L-1)/2; n = 0:1:L-1;
```

```
ImpLoWide = sin(wc2*(n - M + eps))./(pi*(n - M + eps));
ImpLoNarrow = sin(wc1*(n - M + eps))./(pi*(n - M + eps));
ImpBand = ImpLoWide - ImpLoNarrow;
frImpBand = abs(fft(ImpBand,1024));
figure(57); subplot(211); stem(n,ImpBand);
xlabel('(a) Sample'); ylabel('Amplitude')
axis([0 length(ImpBand) 1.2*min(ImpBand) 1.1*max(ImpBand)])
subplot(212); plot([0:1:512]/512,frImpBand(1,1:513));
xlabel(['(b) Frequency, Units of \pi']);
ylabel('Magnitude'); axis([0 1 0 1.1])
```

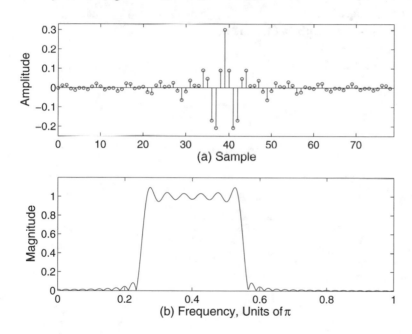

Figure 2.11: (a) Bandpass filter impulse response; (b) Magnitude of DTFT of impulse response in (a).

Example 2.7.   Design a bandstop filter having band edges [0.4, 0.5, 0.7, 0.8], length 71, and using a rectangular window.

To do this, we create a bandpass filter as above having the needed band edges and length, then subtract it from a filter of the same length that passes all frequencies from 0 to π radians. The following code implements this procedure; Fig. 2.12 shows the result of making the call

LVNotchViaLPFSincRectwin(0.45*pi,0.75*pi,71)

```
function LVNotchViaLPFSincRectwin(wc1,wc2,L)
% LVNotchViaLPFSincRectwin(0.45*pi,0.75*pi,71)
M = (L-1)/2; n = 0:1:L-1;
ImpLo2 = sin(wc2*(n - M + eps))./(pi*(n - M + eps));
ImpLo1 = sin(wc1*(n - M + eps))./(pi*(n - M + eps));
ImpBand = ImpLo2 - ImpLo1;
ImpWide = sin(pi*(n - M + eps))./(pi*(n - M + eps));
ImpStop = ImpWide - ImpBand;
frImpStop = abs(fft(ImpStop,1024));
figure(58); subplot(211); stem(n,ImpStop);
xlabel('(a) Sample'); ylabel('Amplitude');
subplot(212); plot([0:1:512]/512, frImpStop(1,1:513));
xlabel(['(b) Frequency, Units of \pi']);
ylabel('Magnitude'); axis([0 1 0 1.2])
```

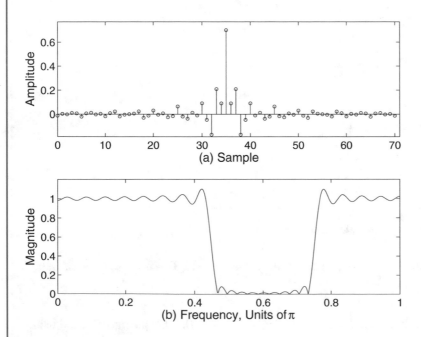

Figure 2.12: (a) Bandstop filter impulse response; (b) Magnitude of DTFT of sequence in (a).

2.5.5   IMPROVING STOPBAND ATTENUATION

The filters designed in the examples above show the high passband ripple and poor stopband attenuation associated with the rectangular window. By choosing different nonrectangular windows, the stopband attenuation can be improved at the expense of a shallower roll-off–but this can be compensated by increasing the filter length.

Example 2.8.   Compare the stopband attenuation, passband ripple, and transition widths of a length-50 lowpass filter, $\omega_c = 0.5\pi$, windowed with rectangular, Hamming, Blackman, and Kaiser (β = 10) windows. Perform the experiment a second time, using L = 101.

We'll present code to design the lowpass filter using the Kaiser(10) window, and illustrate the other windows in Fig. 2.13.

```
function LVLPFViaSincKaiser(wc,L)
% LVLPFViaSincKaiser(0.5*pi,50)
M = (L-1)/2; n = 0:1:L-1;
Imp = sin(wc*(n - M + eps))./(pi*(n - M + eps));
win = kaiser(L,10)'; fr = abs(fft(Imp.*win, 2048));
fr = fr(1,1:fix(length(fr)/2+1));
xvec = [0:1:length(fr)-1]/length(fr);
figure(6); plot(xvec,20*log10(fr+eps))
xlabel(['Freq, Units of \pi'])
ylabel(['Mag, dB']); axis([0 inf -110 10])
```

Figures 2.13 and 2.14 demonstrate that the stopband attenuation for each specific type of window remains the same irrespective of the filter length. Only the transition width is affected by filter length. The transition width is inversely proportional to the filter length; that is, doubling the filter length may generally be expected to halve the transition width. The Kaiser window can, by changing β, be given whatever stopband attenuation level is desired, including values much greater than the Blackman window, for example.

Example 2.9.   Use the guidelines for L to design a lowpass filter, using the Blackman window, having $\omega_p = 0.25\pi$, $\omega_s = 0.32\pi$, A_s = 74 dB, and R_p = 0.1 dB. Verify that the design goals are met.

The following program uses $11\pi/\omega_t$ as the estimate for L, but designs the filter over a range of L from about 90 % to 110 % of the estimated L to ascertain the lowest value of L that meets the design value of A_s. Figure 2.15 shows the result, upon reaching a value of L adequate to meet A_s; note that the final value of L is slightly larger than the estimated value.

```
function LVLPFSincBlackman(wp,ws,Rp,As)
% LVLPFSincBlackman(0.25*pi,0.32*pi,0.1,74)
wc = (wp+ws)/2; wt = ws - wp; limL = ceil(11*pi/wt),
startL = fix(0.9*limL); figure(59); LenFFT = 8192;
for L = startL:1:ceil(1.1*limL); M = (L-1)/2;
n = 0:1:L-1; b = sin(wc*(n - M + eps))./(pi*(n - M + eps));
b = b.*(blackman(L)'); fr = abs(fft(b,LenFFT));
```

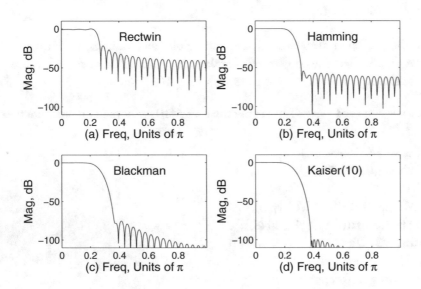

Figure 2.13: (a) DTFT of a length-50 ideal lowpass filter multiplied by a rectangular window; (b) Same, with Hamming window; (c) Same, with Blackman window; (d) Same, with Kaiser window, $\beta = 10$.

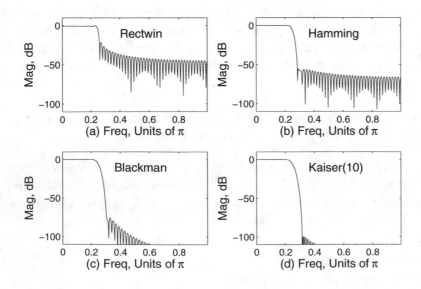

Figure 2.14: (a) DTFT of a length-101 ideal lowpass filter multiplied by a Rectangular window; (b) Same, with Hamming window; (c) Same, with Blackman window; (d) Same, with Kaiser window, $\beta = 10$.

```
fr=fr(1,1:(LenFFT/2+1)); Lfr = length(fr);
PB = fr(1,1:round((wp/pi)*Lfr)); SB = fr(1,round((ws/pi)*Lfr):Lfr);
PBR = -20*log10(min(PB)), SBAtten = -20*log10(max(SB)),
plot([0:1:LenFFT/2]/(LenFFT/2), 20*log10(fr+eps));
xlabel('Frequency, Units of \pi');
ylabel(['Mag, dB']); axis([0 1 -100 5])
text(0.6,-10,['est L = ', num2str(limL)]);
text(0.6,-20,['L = ', num2str(L)]);
text(0.6,-30,['design Rp = ', num2str(Rp)]);
text(0.6,-40,['actual Rp = ', num2str(PBR)]);
text(0.6,-50,['design As = ', num2str(As)]);
text(0.6,-60,['actual As = ', num2str(SBAtten)]);
if SBAtten>-As; break; end; pause(0.25); end
```

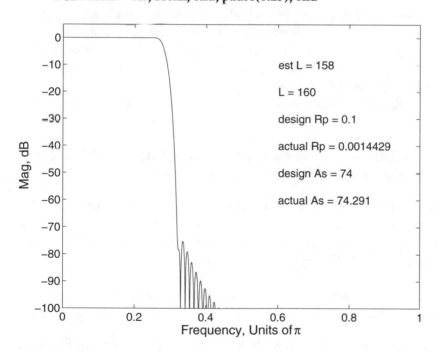

Figure 2.15: The frequency response of a lowpass filter designed using the Blackman window.

2.5.6   MEETING DESIGN SPECIFICATIONS
A basic procedure to design a filter using the windowed ideal lowpass method is as follows:

- Specify the filter design parameters, which consist of filter type, band edges, maximum desired passband ripple, and minimum permissible stopband attenuation.

- Pick a window that can produce the required minimum stopband attenuation.

- Either estimate the value of L needed, or start at a low value of L and proceed incrementally until the minimum stopband attenuation is just met. Inadequate values of L cause the window's transition band to pass into the stopband, thus producing a minimum stopband attenuation above that attainable with the given window. Compare the value of realized passband ripple to the design value.

One of the exercises at the end of this chapter is to generate a script which conforms to the following call syntax:

$$LVxFIRViaWinIdealLPF(FltType, BndEdgeVec, Win, As, Rp) \qquad (2.5)$$

FltType is passed as *1* for lowpass, *2* for highpass, *3* for bandpass, and *4* for bandstop; *BndEdgeVec* is specified as ω_p, ω_s for a lowpass filter, $[\omega_s, \omega_p]$ for a highpass filter, $[\omega_{s1}, \omega_{p1}, \omega_{p2}, \omega_{s2}]$ for a bandpass filter, and $[\omega_{p1}, \omega_{s1}, \omega_{s2}, \omega_{p2}]$ for a bandstop filter, all values in fractions of π, such as 0.3, 0.5, etc, with 1.0 being the Nyquist rate, or π radians. Available window types are *rectwin*, *bartlett*, *hanning*, *hamming*, *blackman*, and *kaiser*. Minimum acceptable stopband attenuation in dB is passed as *As* and maximum acceptable passband ripple in dB as R_p.

The computations and figures for the following examples were generated using a script conforming to function (2.5).

Example 2.10.   Design a bandpass filter having *BndEdgeVec* = [0.4, 0.5, 0.8, 0.9], *Win* = 'kaiser', and *As* = 78. We also require that the maximum passband ripple be 0.1 dB.

The frequency response of the resultant design is displayed along with various parameters (L, β, A_s, R_p) in Fig. 2.16. The script obtained the initial estimate for L and the value of β from the design formulas (2.3) and (2.4). We note that the actual value of passband ripple (R_p) and minimum stopband attenuation are, respectively, 0.0023 dB and 78 dB, which meet the design goals.

Example 2.11.   Compare designs for a lowpass filter with the following design specifications:
A_s = 44, R_p = 0.1, ω_p = 0.45, ω_s = 0.55, using two different windows, Hanning and Kaiser.

The Hanning design is shown in Fig. 2.17, while the Kaiser design is shown in Fig. 2.18. Note that the Kaiser window produces a much shorter filter length for the same design parameters.

Example 2.12.   Compare designs for a bandstop filter having *BndEdgeVec* = [0.4, 0.45, 0.65, 0.7]. A_s = 74, Rp = 0.1, using the Blackman and Kaiser windows.

We again find that the Kaiser window needs a smaller value of L than does the other window (Blackman, in this case) for the same design parameters. The Blackman design is shown in Fig. 2.19, while the Kaiser design is shown in Fig. 2.20.

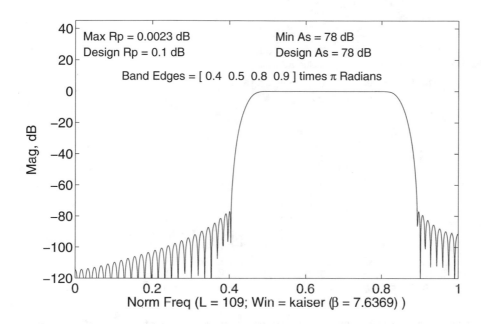

Figure 2.16: The frequency response and various design and realized parameters for a bandpass filter using a Kaiser window.

2.6   FIR DESIGN VIA FREQUENCY SAMPLING

In the previous section, we synthesized a basic lowpass filter impulse response by defining an ideal lowpass frequency response and using the inverse DTFT to determine an equivalent time domain expression which could be evaluated over any desired range of sample index n. We were further able, with some ingenuity, to synthesize highpass, bandpass, and bandstop filters from lowpass filters.

In the Frequency Sampling approach, we again start in the frequency domain with a specification, but instead of using a continuous frequency specification and the inverse DTFT, we take equally spaced samples of the frequency domain specification, and treat them as a DFT. The inverse DFT then yields an impulse response that will result in a filter whose frequency response matches that of the specification exactly at the location of the frequency samples. The filter's frequency response at values of ω lying between the frequency samples will differ from the ideal or continuous frequency response.

Once having specified the desired response, which is a set of amplitudes, one for each correlation frequency possible within the filter length, it is possible to proceed two ways:

- **Convert the specification to a correctly-formatted DFT and perform the IDFT to obtain the impulse response.**

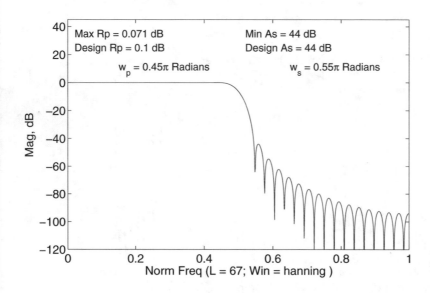

Figure 2.17: Frequency response of lowpass filter designed using a Hanning window with a target A_s of 44 dB and R_p of 0.1dB. The design specifications were met with $L = 67$.

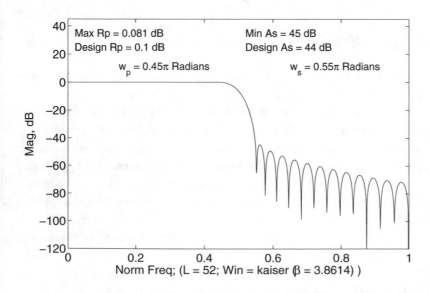

Figure 2.18: Frequency response of a lowpass filter designed using a Kaiser window ($\beta = 3.8614$) with a target A_s of 44 dB and R_p of 0.1dB. The design specifications were met with $L = 52$.

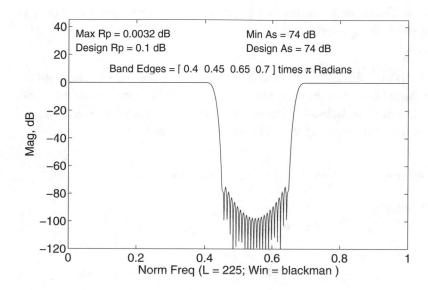

Figure 2.19: Frequency response of a bandstop filter designed using a Blackman window with a target A_s of 74 dB and R_p of 0.1dB. The design specifications were met with $L = 225$.

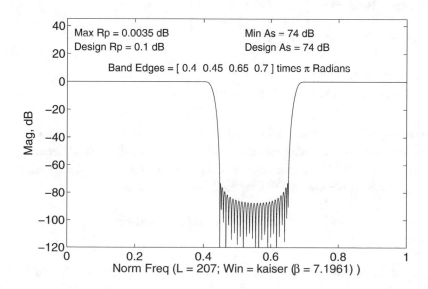

Figure 2.20: Frequency response of a bandstop filter designed using a Kaiser window ($\beta = 7.1961$) with a target A_s of 74 dB and R_p of 0.1dB. The design specifications were met with $L = 207$.

- **Use the desired response values directly in cosine or sine summation formulas to obtain the impulse response. These are formulas that produce the same net result as would be obtained by taking the IDFT. However, no complex arithmetic is involved in using the summation formulas.**

Figure 2.21 shows a filter specification for a length-40 filter. The passband and stopband desired responses are shown as solid lines, and the frequency samples, located at normalized frequencies of $2k/40$ ($k = 0{:}1{:}19$), are marked with circles. The actual frequency response is marked with a dotted line, and the borders of the transition band are marked with vertical dashed lines.

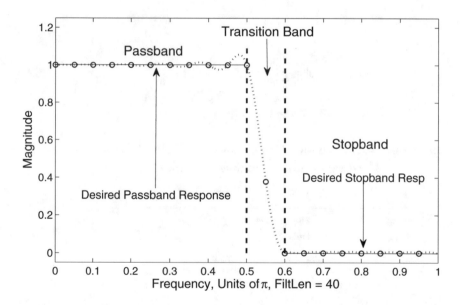

Figure 2.21: A filter frequency specification, with desired passband and stopband responses shown as solid lines. The filter's actual frequency response, shown as a dotted line that passes through all frequency samples (marked with circles), was obtained by performing the DTFT on the filter's impulse response, which was itself obtained by performing the inverse DFT on the frequency samples.

A systematic procedure can be followed to design any standard passband type using the frequency sampling method:

- 1. Define required pass, stop, and transition bands by normalized frequency (1.0 equals Nyquist rate, half the sampling rate).

- 2. Choose a filter type. Available filter types are I, II, III, and IV.

    a. Type I filters can be lowpass, highpass, bandpass, or bandstop.

b. Type II filters can be lowpass or bandpass; highpass and bandstop are prohibited.

c. Type III filters can be bandpass only; lowpass, highpass, and bandstop are prohibited. Hilbert transformers are possible.

d. Type IV filters can be bandpass or highpass; lowpass and bandstop are prohibited. Hilbert transformers and differentiators are possible.

- 3. Estimate the needed filter length L. The longer the filter, the steeper the roll-off. If uncertain, pick a length, design a filter, and, if passband ripple and stopband attenuation are inadequate, increase L. This generally provides additional samples in the transition bands, the values of which, when properly chosen (see discussion below entitled "Improving Stopband Attenuation"), improve stopband attenuation and generally reduce passband ripple as well.

- 4. Choose a frequency allocation method, and then identify the available correlator frequencies. For the first frequency allocation method, you can use summation formulas or the Inverse DFT method; for the second frequency allocation method, formulas are used. The two sets of formulas are as follows:

a) Formulas using integer frequencies such as 0, 1, 2, etc. This method produces normalized correlator frequencies of $2k/L$ (radian frequencies $2\pi k/L$), where L is the filter length, and k runs from 0 to L - 1.

b) Formulas using frequencies that are odd multiples of 0.5, namely, 0.5, 1.5, etc. Normalized frequencies are thus $(2k + 1)/L$ (radian frequencies $(2k + 1)\pi/L$).

Formulas for Type I-IV filters for both types of frequency allocation methods may be found in the Appendices of this book.

Figure 2.22 shows the frequency allocations for a length 19, Type I filter using the first frequency allocation method, while Fig. 2.23 shows similar information using the second frequency allocation method.

- 5. Choose the design formula based on the choices made in Steps 2 and 4 above. In choosing L and frequency allocation method, the goal should generally be to place frequency samples at or very near band edges, with one or more samples in the transition band. It may be necessary to change L, change frequency allocation method, or change band edge specifications to achieve this.

- 6. Assign amplitudes for each correlator frequency, based on the passband, stopband, and transition band assignments made in Step 1 above. Typically, an amplitude of 1.0 is used for passband correlators, 0 for stopband correlators, and if there are transition bands (which there should be in order to achieve good stopband attenuation), values intermediate 0 and 1. There are optimum values for transition band correlators; these have been tabulated in Reference [4] by filter length, frequency allocation method, number of transition samples, and number of consecutive bands having amplitude 1.0 that border on the transition band in question.

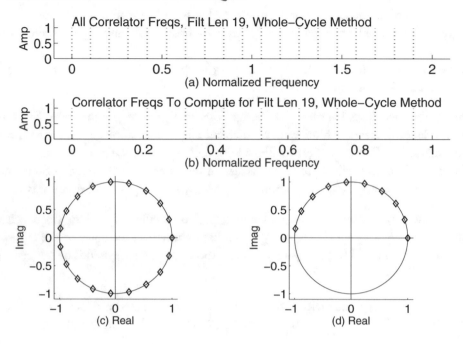

Figure 2.22: (a) All correlator frequencies for Type I filter, length 19, whole-cycle method; (b) The frequencies that must be computed to generate the filter impulse response; (c) Same as (a), but plotted in the complex plane; (d) Same as (b), but plotted in the complex plane.

- 7. Compute the impulse response using the chosen formula and parameters. If the first frequency allocation method is being used, it is straightforward to alternatively use the Inverse DFT method to obtain the impulse response. Often, a window is applied; this tends to reduce ripple and increase stopband attenuation at the expense of roll-off rate. L must be increased accordingly. Note that when a window (other than rectangular) is used, the frequency response will no longer be equal to the specified frequency sample values at the corresponding correlator frequencies.

- 8. Evaluate the frequency response of the resultant filter using the DTFT (usually plotted using a logarithmic scale to show stopband detail). Determine actual values for A_s and R_p.

- 9. Alter filter length L as necessary to attempt to bring the filter's frequency response closer to the design specification.

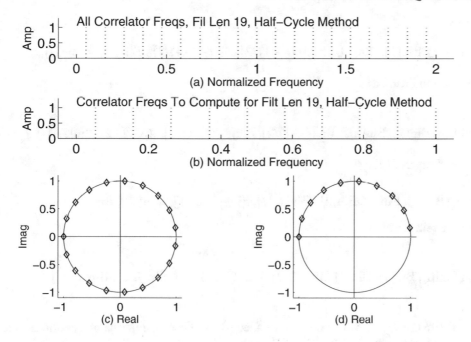

Figure 2.23: (a) All correlator frequencies for Type I filter, length 19, half-cycle method; (b) The frequencies that must be computed to generate the filter impulse response; (c) Same as (a), but plotted in the complex plane; (d) Same as (b), but plotted in the complex plane.

2.6.1 USING THE INVERSE DFT

Setting Bin Amplitudes

We begin by determining the amplitudes for the available correlators. For odd length filters, this amounts to specifying the amplitudes for Bin (or correlator or frequency) 0 and a set of positive bins. Once having this, the negative bin amplitudes may be set to the same value as for corresponding positive bins, i.e., the amplitude for Bin(-1) is the same as for Bin(1).

This produces, for **symmetrically indexed DFTs**, and **odd length** filters, the vector of amplitudes

$$A_k = [\text{Bin}[-(L-1)/2],...\text{Bin}[-2], \text{Bin}[-1], \text{Bin}[0], \text{Bin}[1], \text{Bin}[2],...\text{Bin}[(L-1)/2]]$$

For **even length** filters, there is also a Bin($L/2$):

$$A_k = [\text{Bin}[-L/2 +1],...\text{Bin}[-2], \text{Bin}[-1], \text{Bin}[0], \text{Bin}[1], \text{Bin}[2],...\text{Bin}[L/2]]$$

For **asymmetrically indexed DFTs**, the arrangements are, for **odd length** filters

$$A_k = [\text{Bin}[0], \text{Bin}[1], \text{Bin}[2],...\text{Bin}[(L-1)/2], \text{Bin}[-(L-1)/2],..\text{Bin}[-2], \text{Bin}[-1]]$$

which can be reindexed as

$$A_k = [\text{Bin}[0], \text{Bin}[1], \text{Bin}[2],...\text{Bin}[(L-1)/2], \text{Bin}[L-(L-1)/2],..\text{Bin}[L-2], \text{Bin}[L-1]]$$

and for **even length** filters

$$A_k = [\text{Bin}[0], \text{Bin}[1], \text{Bin}[2],..\text{Bin}[L/2], \text{Bin}[-L/2 +1],..\text{Bin}[-2], \text{Bin}[-1]]$$

which can be reindexed as

$$A_k = [\text{Bin}[0], \text{Bin}[1], \text{Bin}[2],..\text{Bin}[L/2], \text{Bin}[L-L/2 +1],..\text{Bin}[L-2], \text{Bin}[L-1]]$$

Example 2.13.   Write the vector of Bin indices for a length-8 (Type II) filter using symmetric and asymmetric indices. Do the same for a length-9 (Type I) filter. For the asymmetric indices, also write out the vector of actual frequencies.

For symmetric indices and length-8, we get $-L/2 +1:1:L/2 = -3:1:4$. For asymmetric indices, we get $0:1:(L-1) = 0:1:7$, which in terms of frequencies is $[0,1,2,3,4,-3,-2,-1]$.

For the length-9 filter, symmetric indices, we get $-4:1:4$, and for asymmetric indices we get $0:1:8$, which in terms of frequencies is $[0,1,2,3,4,-4,-3,-2,-1]$.

Example 2.14.   Specify the Bin amplitudes for length-8 and length-9 filters that have only Bin[0] and Bin[1] amplitudes set to 1.0 with all other bins set to amplitude 0.

For a length-8, symmetrical DFT, we have indices $-3:1:4$, and set Bins -1, 0, and 1 to amplitude 1, which yields the amplitude vector $[0,0,1,1,1,0,0,0]$. Using asymmetrical indices, we get $[1,1,0,0,0,0,0,1]$. For a length-9 filter, symmetrical DFT, we have indices $-4:1:4$, and therefore we get $[0,0,0,1,1,1,0,0,0]$. Using asymmetrical indices, this becomes $[1,1,0,0,0,0,0,0,1]$.

Setting Bin Phase

This vector of amplitudes must then be multiplied by a phase factor that produces the correct linear phase angles for the various bins. We note that, for **asymmetrically indexed DFTs**, having bins from 0 to $L-1$, the phase angles in radians are given as, for Type I and II filters,

$$\angle H[k] = \begin{cases} -2\pi k M/L & k = 0, 1...M \\ 2\pi(L-k)M/L & k = M+1, M+2, ...L-1 \end{cases}$$

where $\angle H[k]$ is the phase at Bin k and $M = (L-1)/2$. For Type III and IV filters, the phase is

$$\angle H[k] = \begin{cases} (\pm\pi/2) - 2\pi k M/L & k = 0, 1...M \\ (\pm\pi/2) + 2\pi(L-k)M/L & k = M+1, M+2, ...L-1 \end{cases}$$

For **symmetrically indexed DFTs**, having k from $-M$ to M for odd filters and $-L/2 +1$ to $L/2$ for even filters, a single expression for Types I and II is

$$\angle H[k] = -2\pi k((L-1)/2)/L) = -\pi k((L-1)/L)$$

and for Types III and IV

$$\angle H[k] = \pm\pi/2 - 2\pi k((L-1)/2)/L) = \pm\pi/2 - \pi k((L-1)/L)$$

By using $\angle H[k]$ as the argument for the complex exponential, the actual complex-numbered phase angles $H_{ph}[k]$ for the DFT (Types I and II) can be generated:

$$H_{ph}[k] = e^{-j2\pi k((L-1)/2)/L)} = e^{-j\pi k((L-1)/L)} \tag{2.6}$$

Example 2.15.   Design a length-9 (Type I) filter with cutoff approximately equal to 0.5π using the Inverse-DFT method. Compute the impulse response and the frequency response from 0 to π radians. On the frequency response plot, also plot the frequency samples.

We note that the Nyquist limit is $9/2 = 4.5$. By setting correlators 0, 1, and 2 (located at normalized frequencies of 0, 0.222, and 0.444) to amplitude 1, and correlators 3 and 4 (located at normalized frequencies of 0.666 and 0.888) to 0, the cutoff will be between band (or correlators) 2 and 3, i.e., at about sample 2.5, which is $2.5/4.5 = 0.55$. We therefore have the vector of sample amplitudes as

$$A_k = [0,0,1,1,1,1,1,0,0]$$

having bin indices $[-4:1:4]$. We obtain the net DFT by multiplying by

$$H = e^{-j\pi[-4:1:4]((9-1)/9)}$$

Thus, far we have formatted the DFT using symmetric indices. It is necessary to adjust bin location when using DFT/IDFT routines that expect the asymmetric bin arrangement, in which k runs from 0 to $L-1$ rather than from $-(L-1)/2$ to $(L-1)/2$ for odd length filters, or from $-L/2 +1$ to $L/2$ for even length filters, it is necessary to shift the negative bins to the right side of the DFT; this is done in the following script:

```
function LVLPFViaSymm2AsymmIDFT(Ak)
% LVLPFViaSymm2AsymmIDFT([0,0,1,1,1,1,1,0,0]); % odd
```

```
% LVLPFViaSymm2AsymmIDFT([0,0,1,1,1,1,1,0,0,0]) % even
L = length(Ak); if ~(rem(L,2)==0) % odd length filter
M = (L-1)/2; symmDFT = Ak.*exp(-j*pi*[-M:1:M]*(2*M)/L);
LenNegBins = M; NegBins = symmDFT(1,1:LenNegBins);
ZeroPosBins = symmDFT(1,LenNegBins+1:length(symmDFT));
NetDFT = [ZeroPosBins NegBins]; Imp = real(ifft(NetDFT));
k = 0:1:(L-1)/2; else % even length filter
symmDFT = Ak.*exp(-j*pi*[-L/2+1:1:L/2]*(L-1)/L);
LenNegBins = L/2-1; NegBins = symmDFT(1,1:LenNegBins);
ZeroPosBins = symmDFT(1,LenNegBins+1:length(symmDFT));
NetDFT = [ZeroPosBins NegBins]; Imp = real(ifft(NetDFT));
k = 0:1:L/2; end
figure(3); clf; subplot(211); stem([0:1:length(Imp)-1],Imp);
xlabel('Sample'); ylabel('Amplitude')
fr = abs(fft(Imp,1024));subplot(212);
plot([0:1:512]/512,fr(1,1:513));
xlabel('Frequency, Units of \pi')
ylabel('Magnitude'); hold on; for ctr = k;
plot([(2*ctr/L),(2*ctr/L)],[0 1],'b:');
plot([2*ctr/L],abs(ZeroPosBins(1,ctr+1)),'ko');
axis([0 1 -inf inf]); end
```

The result from making the call

$$\text{LVLPFViaSymm2AsymmIDFT}([0,0,1,1,1,1,1,0,0])$$

is shown in Fig. 2.24

Example 2.16.   Design a length-8 (Type I) filter with cutoff approximately equal to 0.5π using the Inverse-DFT method. Compute the impulse response and the frequency response from 0 to π radians. On the frequency response plot, also plot the frequency samples.

We note that the Nyquist limit is 8/2 = 4. By setting correlators 0, 1, and 2 to amplitude 1, and correlator 3 to 0, the cutoff will be between band (or correlators) 2 and 3, i.e., with the cutoff frequency at about 2.5/4 = 0.625. The DFT amplitudes will be

$$A_k = [0,1,1,1,1,1,0,0]$$

having bin indices [-3:1:4]. We obtain the net DFT by multiplying by

$$H_{ph} = e^{-j\pi[-3:1:4](7/8)}$$

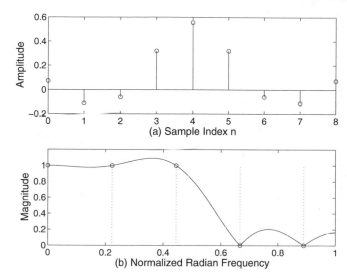

Figure 2.24: (a) Impulse response; (b) Magnitude of DTFT of impulse response in (a), with the frequency samples from the design specification marked with circles.

The computation can be done by the previously presented script, with the result from making the call

LVLPFViaSymm2AsymmIDFT([0,1,1,1,1,1,0,0])

shown in Fig. 2.25. Note that this filter is even in length, and must therefore have a frequency response of zero at the Nyquist limit. To verify this, note that the rightmost value in A_k in the code below is the $L/2$ bin; try changing its amplitude to a nonzero number. You will find that the response at π radians (normalized frequency 1.0) is still zero.

2.6.2 USING COSINE/SINE SUMMATION FORMULAS

Prior to beginning a detailed discussion, the reader should note that Cosine and Sine summation formulas for Types I, II, III, and IV filters can be found in the Appendices to this book.

To start, we will design a Type-II lowpass filter in the following example:

Example 2.17. Design a length-20 (Type II) filter meeting the band requirements shown in Fig. 2.26, where $\omega_p = 0.4\pi$, $\omega_s = 0.5\pi$, $A_s = 40$ dB, and $R_p = 0.5$ dB. The available correlator frequencies are marked for a length 20 (Type II) filter, first frequency allocation method.

The design formula for a Type I or II filter (first frequency allocation method, frequency samples at $2\pi k/L$) is

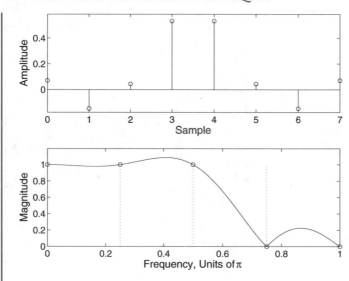

Figure 2.25: (a) Impulse response; (b) Magnitude of DTFT of impulse response in (a), with the frequency samples from the design specification marked with circles.

$$h[n] = \frac{1}{L} \left[A_0 + \sum_{k=1}^{K} 2A_k \cos(2\pi(n - M)k/L) \right]$$

where $K = (L - 1)/2$ for Type I filters and $K = L/2 - 1$ for Type II filters.

For each of the available correlator frequencies $(2\pi k/20)$, we'll specify the desired amplitude of response of the filter. We thus would have for the design amplitude vector

$$A_k = [1, 1, 1, 1, 1, 0, 0, 0, 0, 0]$$

which has k indices 0:1:9 (these would be the same as Bins[0:1:9] if we were using the IDFT method described above).

For both of the following guidelines, *PosBins* does not include Bin 0.

- When the filter is to be odd in length, the filter will have $L = 2*$length(PosBins) +1. When the filter is to be even in length, then $L = 2*$length(PosBins) + 2.

- When starting with a given value of L, for even length filters, length(PosBins) = $(L - 2)/2$, and for odd length filters, length(PosBins) = $(L - 1)/2$.

The following script (see exercises below)

$$WF = LVxFilterViaCosineFormula(Type, Bin0, PosBins)$$

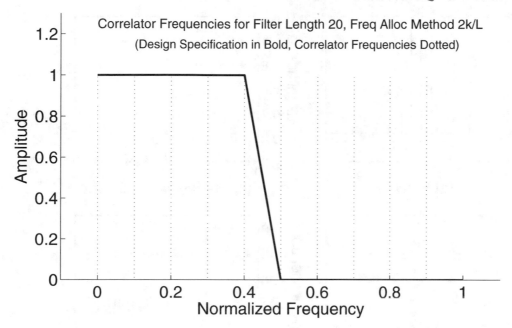

Figure 2.26: Design specification for a lowpass filter.

will compute the impulse and frequency responses for a Type I or II filter, and plot the magnitude of frequency response.

Figure 2.27 shows the result from making the call

WF = LVxFilterViaCosineFormula(2,[1],[1,1,1,1,0,0,0,0,0])

Note that the results from this simple approach are poor in terms of stopband attenuation and passband ripple.

Example 2.18.   Use the script *LVxFilterViaCosineFormula* to design a length-20 bandpass filter having its passband centered around 0.5π radians, and having a passband width of about 0.2π radians.

The length of *PosBins* will be (20-2)/2 = 9. *Bin0* will have an amplitude of 0. The Nyquist limit is 20/2 = 10, and therefore samples 4-6 (*Bin0* being indexed as sample 0) represent normalized frequencies of 0.4, 0.5, and 0.6. The script call will be

WF = LVxFilterViaCosineFormula(2,[0],[0,0,0,1,1,1,0,0,0])

The result of the call is shown in Fig. 2.28.

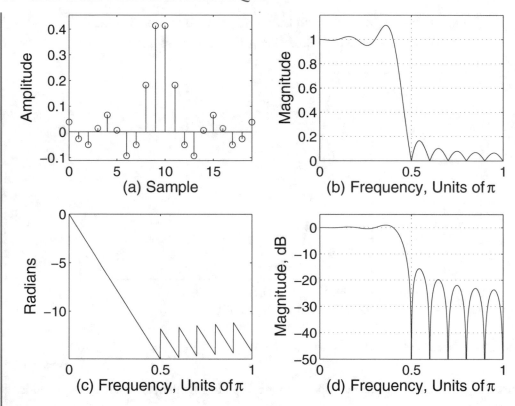

Figure 2.27: (a) Impulse response of a length-20 lowpass filter designed using the cosine summation formula; (b) Magnitude of frequency response of filter at (a); (c) Phase response of same; (d) Magnitude of frequency response in dB.

2.6.3   IMPROVING STOPBAND ATTENUATION

In order to improve stopband attenuation, it is necessary to specify a transition band having within it one or more frequency samples the amplitudes of which may be set to any necessary value to optimize stopband attenuation. This principle is illustrated in the following example.

Example 2.19.   Design a filter having $\omega_p = 0.5\pi$, $\omega_p = 0.6\pi$ and having one frequency sample in the transition band. Programmatically vary the amplitude of the transition band frequency sample and observe the change in stopband attenuation.

Frequency samples at normalized frequencies of 0.5, 0.55, and 0.6 will define the limits of passband and stopband, with a single sample in the transition band as desired. The normalized sample frequencies are, for an odd length filter

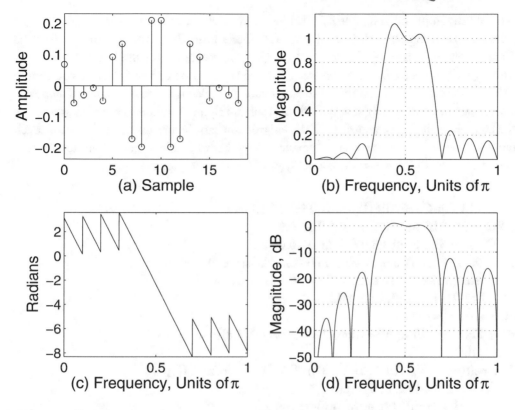

Figure 2.28: (a) The impulse response of a length-20 bandpass filter designed using the cosine summation formula; (b) Magnitude of frequency response of filter at (a); (c) Phase Response of filter at (a); (d) Magnitude of frequency response in dB.

$$2([0:1:M])/L$$

where $M = (L-1)/2$, or

$$2([0:1:(L/2-1)])/L$$

for an even length filter. The sample spacing in radian frequency is $2\pi/L$, or just $2/L$ for normalized frequency.

Since the samples must be equally spaced at normalized frequency interval $2/L$, we get a filter length of $L = 2/0.05 = 40$. The frequency specification must cover Bins 0 to $L/2-1 = 19$ (for an even length filter, Bin $L/2$ is always 0 and is thus not specified), and Bins -1 to -19 having the same amplitude as Bins 1 through 19. The passband for positive Bins runs from Bin 0 to Bin 10. This is followed by the variable T, representing the amplitude of the transition band sample, and a number of zeros to fill out the stopband to sample $L/2-1$. The code below uses the cosine summation formula to obtain the impulse response of the filter for each Ak. For an initial test run, the limits of T (the transition band frequency sample) should be set to give a first estimate of the best value of T by stepping through all the possible values between 1.0 and 0 with a decrement of 0.01. You should find that 0.39 or thereabouts appears to be the best value. Note also that the best value of stopband attenuation appears to occur when the amplitude of stopband ripple is equalized as much as possible (later in the chapter we'll discuss equiripple FIR design). You can then rerun the code with a much reduced range of possible values for T and a smaller decrement to obtain a more accurate estimate. For the second run, try **Thigh = 0.4, Tlow = 0.38, and Dec = 0.0005,** for example.

```
function LVOptCoeffLPF(L,wp,ws,THi,TLo,Dec)
% LVOptCoeffLPF(40,0.5,0.6,0.5,0.35,0.01)
% LVOptCoeffLPF(40,0.5,0.6,0.4,0.38,0.0005)
Dec = -abs(Dec); noComp = ceil((THi-TLo)/abs(Dec)); ctr = 0;
TSB = zeros(noComp,3); LenFFT = 4096; if rem(L,2)==0;
limK = L/2-1; else; limK = (L-1)/2; end;
n = 0:1:L-1; M=(L-1)/2;
for T = THi:Dec:TLo; Ak = [ones(1,11), T, zeros(1,8)];
LA=length(Ak);
WF=(cos(((n-M)')*[0:1:LA-1]*2*pi/L))*([Ak(1),2*Ak(2:LA)]');
WF = WF'; Imp = WF/L; figure(3);
subplot(211); stem([0:1:length(Imp)-1],Imp);
xlabel('n'); ylabel('Amplitude');
fr = abs(fft(Imp,LenFFT));
fr = fr(1,1:LenFFT/2+1); fr = fr/(max(abs(fr)));
Lfr = length(fr); SB = fr(1,round(ws*(Lfr-1)):(Lfr-1));
PB = fr(1,1:round((wp/pi)*(Lfr-1))); subplot(212);
SBAt = -20*log10(max(SB) +eps);
PBR = -20*log10(min(PB)); ctr = ctr + 1;
TSB(ctr,1)=T; TSB(ctr,2)=SBAt; TSB(ctr,3)=PBR;
plot([0:1:Lfr-1]/(Lfr-1),20*log10(fr+eps));
strSBA = num2str(SBAt);
xlabel(['Norm.Freq.(T = ',num2str(T),' and As = ',strSBA,')'])
ylabel('Magnitude'); axis([0 1 -70 10]); pause(0.01);
end; sc = TSB(:,2); bestSB = max(sc),
bestSBind = find(sc==bestSB); bestT = TSB(bestSBind,1),
finalPBRipple = TSB(bestSBind,3)
```

For the example above, we would make the call

LVxFilterViaCosineFormula(2,1,...
[ones(1,10), 0.387, zeros(1,8)])

which results in Fig. 2.29.

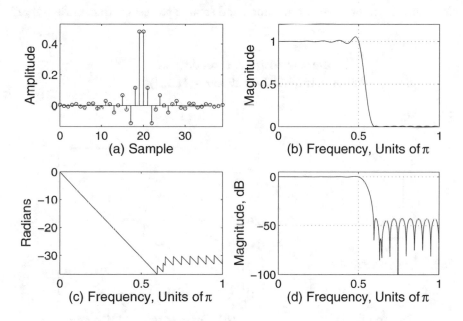

Figure 2.29: (a) Impulse response of a length-40 lowpass filter having a single optimized transition band sample; (b) Magnitude of frequency response of same; (c) Phase response of same; (d) Magnitude of frequency response, in dB.

To improve stopband attenuation further, the transition band can include additional samples, each of which will have an optimum value. Optimum values have been computed for a number of different filter types and lengths, and are found in Reference [4]. While precise determination of optimum values can require optimization algorithms, it is possible to obtain reasonable estimates using simple empirical search techniques.

Example 2.20. Redesign the filter for the previous example, which was a lowpass filter having ω_p = 0.5, ω_s = 0.6. For the redesign, use two optimized transition band samples, and then perform a second redesign using three optimized transition band samples.

The new filter length for two transition band samples can be determined by noting that the samples in or bordering on the transition band should be located at normalized frequencies of 0.5,

0.533, 0.566, and 0.6 which suggests the minimum spacing as $2\pi/L = 0.0333\pi$, which implies that $L = 2/0.0333 = 60$, with

$$A_k = [\textbf{ones(1,16)}, T_1, T_2, \textbf{zeros(1,12)}]$$

The values $T[1] = 0.592$ and $T[2] = 0.109$, which were found by a simple search technique, provide acceptable optimization of the two transition band sample values; we thus make the call

LVxFilterViaCosineFormula(2,1,...
[ones(1,15),0.592,0.109,zeros(1,12)])

which results in Fig. 2.30.

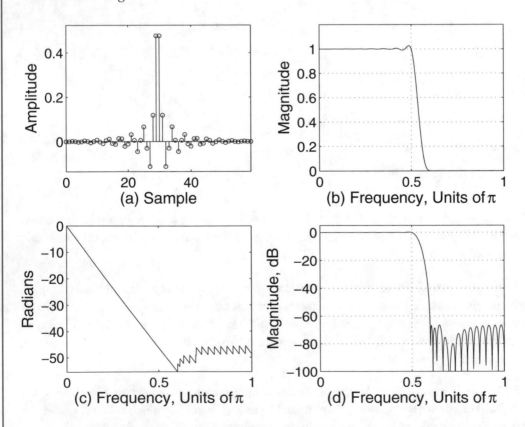

(a) Sample

(b) Frequency, Units of π

(c) Frequency, Units of π

(d) Frequency, Units of π

Figure 2.30: Frequency response of a length-60 LPF with $\omega_p = 0.5\pi$ and $\omega_s = 0.6\pi$, using two transition coefficients with approximately optimum values to achieve the maximum stopband attenuation.

2.6.4 FILTERS OTHER THAN LOWPASS

Recall that in designing filters using the window technique, we started with an Ideal LPF, truncated it to a desired length, and picked a window to achieve a certain stopband attenuation. Filters other than lowpass were synthesized by manipulating one or more lowpass types to generate a different type such as highpass, bandpass, and bandstop. Designing high-quality transition-band-optimized filters via Frequency Sampling, for filter types other than lowpass, in general, requires an optimization program or algorithm. Reference [4] does, however, describe creating a bandpass filter by rotating the A_k samples, optimized for a lowpass response, to a new center frequency, and duplicating the lowpass passband and transition band samples symmetrically on both sides of the new frequency, leaving stopbands of appropriate length on each side of the new pass- and transition bands.

Example 2.21. Convert the lowpass frequency sample vector Ak = [$ones$(1, 6), 0.5943, 0.109, $zeros$(1, 25)] to one suitable for a bandpass filter. Display a magnitude plot for the original lowpass filter and the new bandpass filter.

To do this, a new bandpass characteristic is made consisting of the original lowpass coefficients preceded by a left-right reversed version–this creates a symmetrical bandpass, the center frequency of which is somewhere between 0 and 1.0, normalized frequency. Note that to create a bandpass passband of width PB radians, it is necessary to determine a set of lowpass frequency samples for a passband of $PB/2$ radians. The following call follows the procedure just mentioned. Figure 2.31 shows the original lowpass characteristic, while Fig. 2.32 shows the bandpass characteristic derived therefrom.

LVxFilterViaCosineFormula(1,0,[zeros(1,8),...
0.109,0.5943,ones(1,12),0.5943,0.109,zeros(1,8)])

Note that the stopband attenuation of the bandpass filter is poorer than that achieved by the original set of frequency samples used to generate a lowpass filter. This is expected, as noted in Reference [4]; if better results are needed, optimization would have to be performed directly on the set of bandpass frequency samples. A better, more optimized set of transition bands for the same bandpass filter, given in the following call (the results from which are shown in Fig. 2.33), improves the stopband attenuation significantly.

LVxFilterViaCosineFormula(1,0,[zeros(1,8),0.0875,...
0.5446,ones(1,12),0.5446,0.0875,zeros(1,8)])

The call

LVxFilterViaCosineFormula(1,0,[zeros(1,15),...
0.0165,0.2042,0.6765,ones(1,15)])

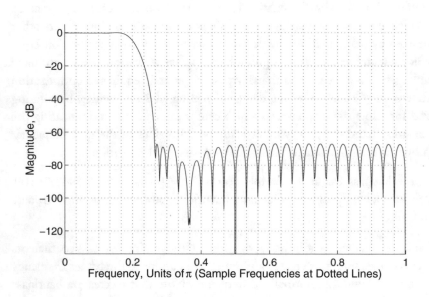

Figure 2.31: Frequency response of a lowpass filter created from the frequency sample vector [ones(1,6), 0.5943, 0.109, zeros(1,25)]. Note minimum A_s of about 67 dB.

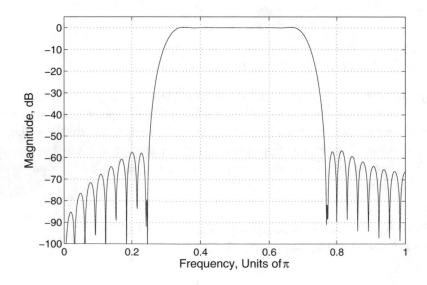

Figure 2.32: Frequency response of a bandpass filter created from the lowpass frequency sample vector [ones(1,6), 0.5943, 0.109, zeros(1,25)]. Note minimum A_s of about 57 dB.

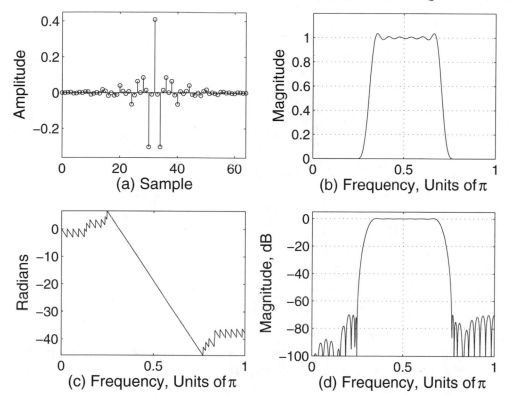

Figure 2.33: (a) Impulse response of a length-65 BPF with three (approximately) optimized transition band coefficients; (b) Magnitude of frequency response of same; (c) Phase response of same; (d) Frequency response in dB.

produces Fig. 2.34, a nearly-optimized highpass filter with three transition band coefficients.

Several VIs that illustrate this specific example, and allow the user to vary the three transition coefficients manually while viewing the resultant frequency response are

$$DemoHPFOptimizeXitionBandsVI$$

$$DemoHPFOptimizeXitionBandsPrecVI$$

The former VI allows coefficient entry using virtual slider controls, while the latter allows precise numerical entry. Figure 2.35 shows an example of the latter VI.

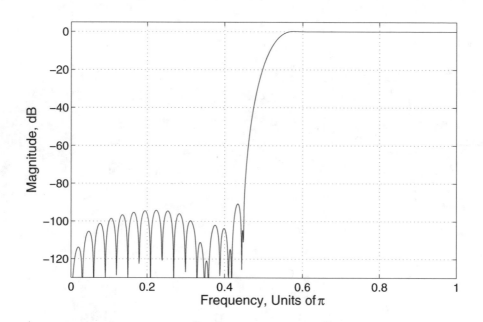

Figure 2.34: A highpass filter having three transition samples with approximate values of T = [0.0165, 0.2042, 0.6765] and stopband attenuation of more than 90 dB.

2.6.5   HILBERT TRANSFORMERS

The Hilbert transformer is a special type of filter the purpose of which is to shift the phase of all the frequencies (except DC and the Nyquist rate) in a signal by 90 degrees ($\pi/2$ radians), leaving the amplitudes untouched. Thus, the ideal Hilbert Transformer has a frequency response

$$H(\omega) = \begin{cases} -j & 0 < \omega < \pi \\ +j & -\pi < \omega < 0 \end{cases} \tag{2.7}$$

The Hilbert transform is useful in communications, for example, to create single sideband signals and the like.

Two approaches to Hilbert-transforming a test waveform are:

1. Frequency Domain: Compute the DFT of the test sequence, multiply it by a frequency domain representation or mask of a Hilbert Transformer, then compute the Inverse DFT to obtain the Hilbert Transform.

2. Time Domain: Design a Hilbert Transform impulse response, and convolve the time domain test signal with it, as with any time domain filter.

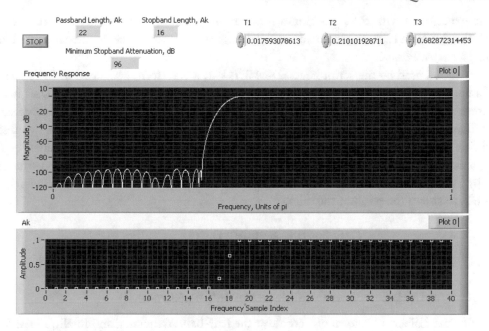

Figure 2.35: A VI allowing precise entry of transition band coefficients T1-T3 for a length-81 FIR HPF having Ak = [zeros(1,16),T1,T2,T3,ones(1,22)], showing good estimates for the three coefficients, resulting in about 96 dB maximum stopband attenuation.

In Frequency Domain

Several methods exist to generate a frequency domain representation of a Hilbert Transformer which will yield the Hilbert Transform of a signal when the DFT of the signal is multiplied by the frequency domain Hilbert mask and the product used to compute the inverse DFT.

- A first method is to generate a frequency domain mask consisting of the value 0 for bins 0 and $N/2$, -j for the remaining positive bins, and +j for negative bins. This method is referred to hereafter as the *All-Imaginary Hilbert Mask* (or Method). The mask is multiplied by the signal's DFT, the inverse DFT is taken, and the real part of the result is the Hilbert Transform.

- A second method is to generate a mask consisting of the value 1 for bins 0 and N/2, 2 for the remaining positive bins, and 0 for all negative bins. This method is referred to hereafter as the *All-Real Hilbert Mask*. The mask is multiplied by the signal's DFT, and the imaginary part of the inverse DFT of the product is the Hilbert Transform, while the real part is the original signal.

The script

$$LVHilbertPhaseShift(DestWF, UserWF, SR)$$

allows you to experiment with the Hilbert transform as computed using frequency domain techniques. Both the All-Real Hilbert Mask and the All-Imaginary Hilbert Mask are demonstrated in the function *LVHilbertPhaseShift*.

The test signal used by the script is, for *Dest W F* = 1, a waveform generated as the superposition of a harmonic series of cosines inversely weighted by harmonic number, or, for *Dest W F* = 2, a waveform generated as the superposition of an odd-harmonic-only series of cosines inversely weighted by harmonic number, or, for *Dest W F* = 3, a user-supplied test signal. The first two test signals, had sine waves been used instead of cosine waves as the basis functions, would have been, respectively, a sawtooth wave and a square wave. The script shifts the phases of the constituent frequencies by 90 degrees and displays the result, using both frequency domain mask methods.

Figure 2.36 shows the result of the call

$$\textbf{LVHilbertPhaseShift(2,[],512)}$$

while Fig. 2.37 shows the result from making the call

$$\textbf{LVHilbertPhaseShift(3,[cos(2*pi*16*[0:1:63]/64)],[])}$$

which computes the Hilbert transform of a cosine at the half-band frequency, and finally, Fig. 2.38 shows the result of the call

$$\textbf{LVHilbertPhaseShift(3,[cos(2*pi*32*[0:1:63]/64)],[])}$$

which computes the Hilbert transform of a cosine at the Nyquist rate.

Example 2.22.   Compute the Hilbert Transform of a one-cycle, four-sample cosine wave using an All-Real FD Mask, using the functions fft and ifft.

We make the following call (note that cos(2*pi*(0:1:3)/4) = [1,0,-1,0]):

$$\textbf{s = [1,0,-1,0]; h = imag(ifft(fft(s).*[1,2,1,0]))}$$

which yields [0,1,0,-1], which is a four-sample sine wave.

We can use a recursive procedure which starts with the original signal and conducts as many 90-degree phase shifts as desired. The following Command Line call returns the original signal after four 90 degree phase shifts; the original signal is introduced as $h = [1,0,-1,0]$ for injection into the recursive loop. Each intermediate result is printed on the Command Line for reference. The outputs will be seen to form the succession of cosine, sine, -cosine, -sine, and cosine again after the fourth 90 degree phase shift.

$$\textbf{h = [1,0,-1,0], for ctr = 1:1:4;s = h;}$$
$$\textbf{h = imag(ifft(fft(s).*[1,2,1,0])),end}$$

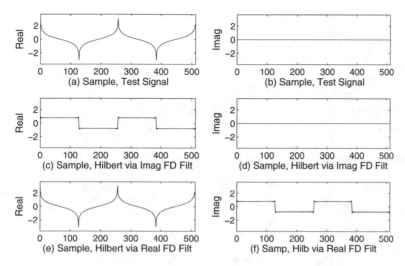

Figure 2.36: (a) Real part of test waveform, sum of odd harmonic cosines with amplitudes inversely proportional to harmonic number; (b) Imaginary part of same; (c) and (d) The real and imaginary parts of the IDFT of the product of the DFT of the waveform at (a) & (b) and an All-Imaginary FD Hilbert mask; (e) and (f) The real and imaginary parts of the IDFT of the product of the DFT of the waveform at (a) & (b) and an All-Real FD Hilbert mask.

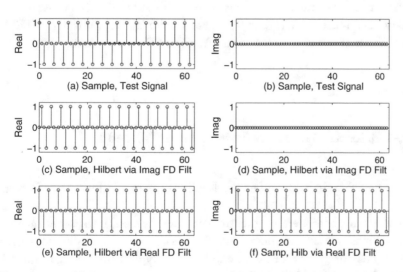

Figure 2.37: (a) Real part of test waveform, a cosine at the halfband frequency ($\pi/2$ radians); (b) Imaginary part of same; (c) and (d) The real and imaginary parts of the IDFT of the product of the DFT of the waveform at (a) & (b) and an All-Imaginary FD Hilbert mask; (e) and (f) The real and imaginary parts of the IDFT of the product of the DFT of the waveform at (a) & (b) and an All-Real FD Hilbert mask.

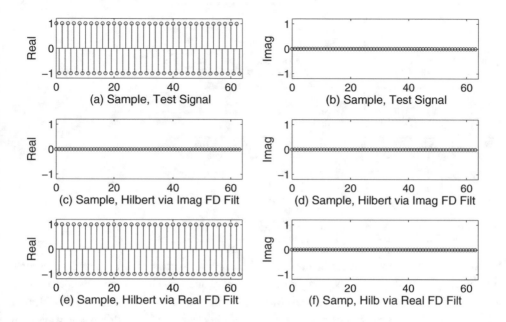

Figure 2.38: (a) Real part of test waveform, a cosine at the Nyquist rate; (b) Imaginary part of same; (c) and (d) The real and imaginary parts of the IDFT of the product of the DFT of the waveform at (a) & (b) and an All-Imaginary FD Hilbert mask; (e) and (f) The real and imaginary parts of the IDFT of the product of the DFT of the waveform at (a) & (b) and an All-Real FD Hilbert mask.

A call using the function *hilbert* is:

h = [1,0,-1,0], for ctr = 1:1:4; s = h; h = imag(hilbert(s)), end

In Time Domain

There are two approaches to employing time domain methods to implement a Hilbert Transformer.

- The first method (probably more of academic rather than practical interest) is to use the inverse DFT on either of the FD masks discussed above. This produces an impulse response suitable for circular convolution (not linear convolution as employed by typical time domain filters) with a signal of the same length to yield the Hilbert Transform.

- The second method is to generate an impulse response suitable for use as a Hilbert Transformer in linear convolution. This can be done using frequency sampling methods or explicit formulas that define the values of the impulse response. One example of the latter is the formula

$$h[n] = \begin{cases} 2\sin^2(\pi n/2)/n\pi & n \neq 0 \\ 0 & n = 0 \end{cases} \tag{2.8}$$

In practice, the maximum magnitude of n will be limited and so the resultant impulse response is only an approximation to the ideal magnitude response. Figure 2.39 shows a length-127 approximation to an ideal Hilbert Transformer, which was generated using (2.8).

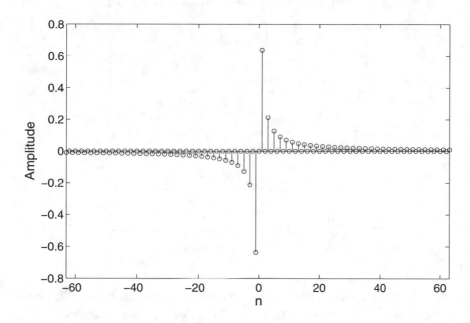

Figure 2.39: A time domain Hilbert Transformer suitable for linear convolution with a signal.

- Hilbert Transformer impulse responses can be generated using the frequency sampling method, either via IDFT or sine summation formulas.

The script (see exercises below)

$$WF = LVxFIRViaWholeSines(AkPos, AkLOver2, L)$$

provides the impulse response, frequency and phase responses, zero plot, etc. for a Type III or IV filter specified by the arguments for Ak, which are partitioned into $AkPos$, the positive frequencies (starting at frequency 1), and $AkLOver2$, which is passed as 0 or the empty matrix for a Type III filter, or a desired amplitude of response for a Type IV filter. The desired filter length is L, and the impulse response is output as WF. The call (in which certain frequency sample amplitudes have been set to values less than 1.0 to optimize the magnitude of the response)

WF = LVxFIRViaWholeSines([0.75,ones(1,23),0.55],[],51)

results in Fig. 2.40.

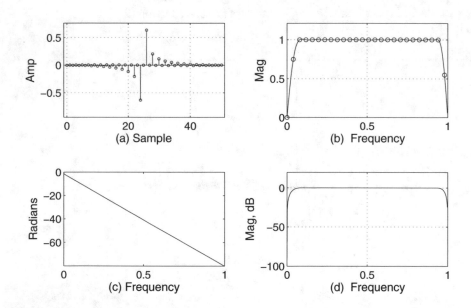

Figure 2.40: (a) Length-51 Hilbert transformer designed by the Frequency-Sampling-IDFT method; (b) Magnitude response of the Hilbert transformer, with DFT samples plotted as circles; (c) Phase response of Hilbert transformer; (d) Magnitude response in dB.

• A direct time domain formula for constructing a Hilbert transformer is as follows:

$$h[n] = \begin{cases} 2\sin^2(\pi(n-M)/2)/n\pi & n-M \neq 0 \\ 0 & n-M = 0 \end{cases} \tag{2.9}$$

where $n = 0:1:L\text{-}1$ and $M = (L\text{-}1)/2$.

The script (see exercises below)

$$LVxHilbertViaConvolution(TestSeqLen, ...$$
$$TestWaveType, FilterLen, TestSigFreq, FDMethod)$$

generates a waveform from cosines having the amplitude structure of a user-selectable sawtooth or square wave, and then phase shifts it by convolving the waveform with a time domain Hilbert

Transformer made using (1) a direct-synthesis time domain formula according to Eq. (2.9) and (2) the Inverse DFT. The first time domain impulse response has length *FilterLen* and is used in linear convolution with the test signal. The second time domain impulse response is generated by the IDFT, has the same length as the test signal, and is suitable for, and used in, circular convolution with the test signal rather than linear convolution. A first window displays the test signal, the directly-generated time domain Hilbert Transformer according to Eq. (2.9), and the convolution of the test signal and the time domain Hilbert Transformer. A second window shows the test signal (real and imaginary parts), a time domain Hilbert Transformer (suitable for circular convolution) of the same length as the test signal, constructed from one of two different frequency domain masks. A circular convolution is performed between the test signal and this second time domain Hilbert Transformer, and the result is displayed in the window.

A typical call is:

LVxHilbertViaConvolution(64,1,19,5,2,0)

which calls for (1) a test signal sequence length of 64, (2) test signal as a sawtooth wave, (3) the time-domain Hilbert Transformer for linear convolution as a length 19 linear-phase FIR, (4) a frequency of 5 cycles (within 64 samples) for the test signal, and (5) an All-Imaginary FD Hilbert Mask method to implement a time-domain Hilbert Transformer for circular convolution with the test signal, and (6) no windowing of the time-domain Hilbert Transformer for linear convolution. Note that when an All-Real FD mask (which is not conjugate symmetric) is used, the equivalent time domain impulse response is complex, whereas when an All-Imaginary FD Mask (which is conjugate symmetric) is used, the equivalent time domain impulse response is real.

The above call results in Figs. 2.41 and 2.42.

Example 2.23. Use Hilbert transforms to create a single sideband signal.

Modulating a carrier wave C with another (usually much lower frequency) frequency S_M creates sum and difference frequencies called sidebands. It is possible, using Hilbert transforms, to eliminate one of these sidebands. Since only one sideband is necessary to transmit all of the intelligence, the effective power increases dramatically over double sideband transmission, in which power goes into both sidebands. The Single Sideband (SSB) mode of radio transmission is popular for this reason.

Multiplying a carrier wave C by a modulating signal S_M results in a double sideband signal:

$$S_{DSB} = C S_M$$

The following expression will eliminate one of the sidebands.

$$S_{SSB} = C S_M \pm (hilbert(C))(hilbert(S_M)) \tag{2.10}$$

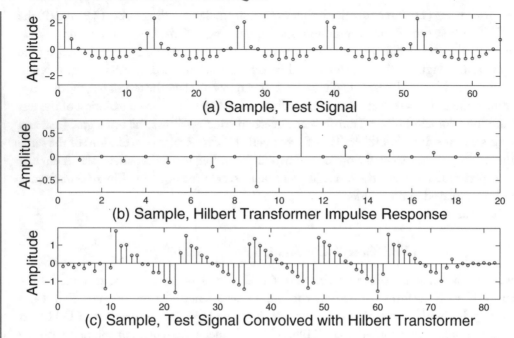

Figure 2.41: (a) Test signal; (b) Hilbert Transformer suitable for linear convolution; (c) Convolution of signals at (a) and (b).

This can be demonstrated, without loss of generality (since F_C, F_S, θ and ϕ are arbitrary), by couching it in a more specific manner as

$$S_{SSB} = \cos(2\pi t F_C + \theta)\cos(2\pi t F_M + \phi) \pm \sin(2\pi t F_C + \theta)\sin(2\pi t F_M + \phi)$$

and then substituting the Euler identity for the trigonometric terms, where, for example, in general

$$\cos(\alpha) = (\exp(j\alpha) + \exp(-j\alpha))/2$$

and

$$\sin(\alpha) = (\exp(j\alpha) - \exp(-j\alpha))/(2j)$$

After substituting the identities for all terms and simplifying, one of the sidebands vanishes.

To demonstrate this, we generate a double sideband signal, then follow that with the single sideband signal and plot the spectra of both. Using $Sb = 0$ leaves the lower sideband, using $Sb = 1$ leaves the upper sideband. Here we have chosen the phase angle for the carrier to be $\pi/2$, and that of the signal as $\pi/6$. You can arbitrarily change the phase angles to verify the generality.

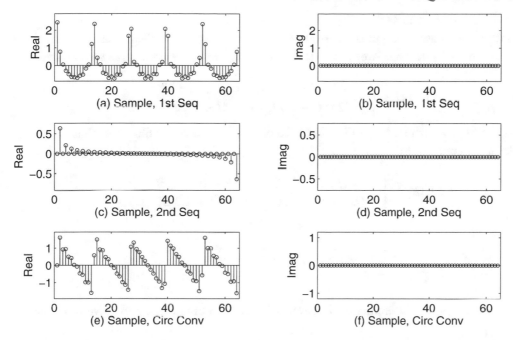

Figure 2.42: (a) and (b): Real and imaginary parts of the test signal; (c) and (d): Real and imaginary parts of TD Hilbert Mask, obtained as the IDFT of an All-Real FD Hilbert Mask; (e) and (f): Real and imaginary parts of circular convolution of test signal and TD Hilbert Mask.

```
function LVSsbDsb(Fc,Fa,Phi1,Phi2,Sb,SR)
% LVSsbDsb(100,20,pi/2,pi/6,0,1000)
phi1 = pi/2; phi2 = pi/6; t = 0:1/SR:1-1/SR;
argsC = 2*pi*t + phi1; argsS = 2*pi*t + phi2;
dsb = cos(argsC*Fc).*cos(argsS*Fa);
figure(95); subplot(2,1,1);
plot(2*t,abs(fft(dsb))); xlabel('Freq, Units of \pi')
ylabel('Mag, DSB Signal')
if Sb==0; ssb = dsb + sin(argsC*Fc).*sin(argsS*Fa);
else; ssb = dsb - sin(argsC*Fc).*sin(argsS*Fa); end
subplot(2,1,2); plot(2*t,abs(fft(ssb)))
xlabel('Freq, Units of \pi'); ylabel('Mag, SSB Signal')
```

For actual signals, which may occupy a band of frequencies, it necessary to use Eq. (2.10) to generate a single sideband signal since there would be many frequencies of arbitrary phase and amplitude present.

2.6.6 DIFFERENTIATORS

A filter that produces as its output the derivative of the input signal is called a differentiator. The frequency response of a differentiator increases linearly with frequency; its sampled response is

$$jHr[k] = \begin{cases} j2\pi k/L & k = 0:1:M \\ -j2\pi(L-k)/L & k = M+1:L-1 \end{cases}$$

where L is the filter length and $M = (L-1)/2$. To obtain a linear phase filter, the phase angles of the samples must conform to

$$\angle H[k] = \begin{cases} -\pi k(L-1)/L & k = 0:1:M \\ \pi(L-k)(L-1)/L & k = M+1:L-1 \end{cases}$$

with

$$H[k] = jHr[k]e^{j\angle H[k]}$$

Example 2.24. Design a differentiator of length 24. Use it to obtain the derivative of a triangle wave.

A Type IV filter is suitable for a differentiator since the required response at the Nyquist limit is nonzero.

```
function LVDifferentiatorLen24
Bins = pi*( [ 0, j*(1:1:11)*2/24, j*12*2/24, j*(-11:1:-1)*2/24 ] )
L = length(Bins); M = (L-1)/2;
kZeroPosBns = 0:1:ceil(M); kNegBns = ceil(M)+1:1:L-1;
angVec = [-2*pi*kZeroPosBns*M/L,2*pi*(L-kNegBns)*M/L];
PhaseFac = exp(j*angVec); NetBins = Bins.*PhaseFac;
Imp = real(ifft(NetBins))
figure(98); stem(Imp); xlabel('Sample'); ylabel('Amplitude')
```

Code that uses the sine summation formulas to construct a length-24 impulse response is

```
function LVDifferL24ViaSineSumm
AkPos = -[1:1:11]*pi/12; AkLOver2 = [-12]*pi/12;
L = 24; M = (L-1)/2; Imp = zeros(1,L); n = 0:1:L-1;
Ak = [AkPos,AkLOver2]; limK = L/2;
for k = 1:1:limK
if (k==limK); C = 1; else; C = 2; end;
Imp = Imp + C*Ak(k)*sin(2*pi*(n-M)*k/L); end;
Imp = Imp/L; figure(109); stem(Imp);
xlabel('Sample'); ylabel('Amplitude')
```

The derivative of a triangle wave is a squarewave, as shown by Fig. 2.43, in which a test triangle waveform appears in plot (a), followed in plot (b) by the convolution of the test waveform with a differentiator. The writing of a script to generate a suitable test wave and differentiator and convolve them to result in a squarewave output is covered in the exercises below.

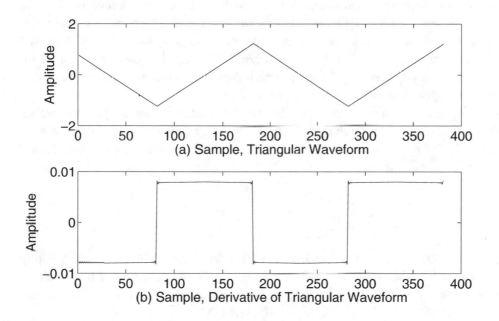

Figure 2.43: (a) Input signal, a Triangle waveform; (b) Derivative of signal in (a), delayed/offset by the filter delay.

2.7   OPTIMIZED FILTER DESIGN

Earlier in the chapter, we designed linear phase filters using the windowing and frequency sampling methods. These methods are straightforward and generally allow direct control over only stopband attenuation. Another approach, more advanced, is to design the filter such that the amplitudes of ripples in the frequency response are equalized. This results in a filter of the lowest order (shortest length) to achieve a given set of criteria; both passband ripple and stopband attenuation can be incorporated into the design.

2.7.1   EQUIRIPPLE DESIGN

The windowed-ideal-response and frequency-sampling methods of FIR design, while effective, do not allow precise control over ripple in both the passband and stopband; nor, in general, is the shortest possible filter length achieved for a given design. When the ripple in the frequency response is equalized, the minimum filter length required can be achieved. An interesting experiment that suggests this can be conducted by running the script

LVFrqSmpLPFOptOneCoeff(0.5,0.6,40,0.40,0.37,-0.001,[]);

Press any key for the next computation. The best stopband attenuation will occur when the transition band coefficient T equals 0.387. Note that the stopband ripple makes its best approach to being equalized at this point. Figure 2.44 shows the result for values of T = to 0.5, 0.387 (the ideal value), and 0.33. Passband ripple for T = 0.387 was 0.67 dB. Figure 2.45 shows the result for the same bandlimits, but with a true equiripple design which does not exceed passband ripple of 0.69 dB. In this case, the stopband attenuation is in excess of 50 dB.

For the FIR design techniques we have seen thus far, the frequency response typically contains ripples through the entire spectrum. Such ripples are not equally spaced, but rather tend to be more closely spaced the closer they are to the edges of a transition band. The number of ripples is limited and generally equal to $R + 2$, where

$$R = \begin{cases} (L-1)/2 & \text{Type-I} \\ (L-3)/2 & \text{Type-III} \\ L/2 - 1 & \text{Type-II, IV} \end{cases}$$

Some filters, known as extra ripple filters, can have $R + 3$ ripples. Figure 2.46 shows an equiripple lowpass filter of length 9 having $\omega_p = 0.4\pi$ and $\omega_s = 0.6\pi$. The number of extremal frequencies in the frequency response is $(9-1)/2 + 2 = 6$. In the case shown in the figure, the extremal frequencies are at normalized frequencies 0, 0.2625, 0.4, 0.6, 0.7375, and 1.0.

2.7.2   DESIGN GOAL

Earlier in the chapter, we noted that, for a general FIR design (for a lowpass filter), the passband frequency response design specification is

$$1 - \delta_1 \leq H_r(\omega) \leq 1 + \delta_1$$

for $\omega < \omega_p$, and for the stopband

$$-\delta_2 \leq H_r(\omega) \leq \delta_2$$

for $\omega > \omega_s$. Thus the maximum passband ripple is δ_1 and the maximum stopband ripple is δ_2.

If we define the FIR's actual response as $P(\omega)$, then the error signal $E(\omega)$ is

$$E(\omega) = H_{dr}(\omega) - P(\omega)$$

The design goal is to determine filter coefficients that will minimize the maximum magnitude of $E(\omega)$ over a filter's passband(s) and stopband(s). This is called a **minimax** problem, that is, minimizing the maximum magnitude of a quantity such as $E(\omega)$.

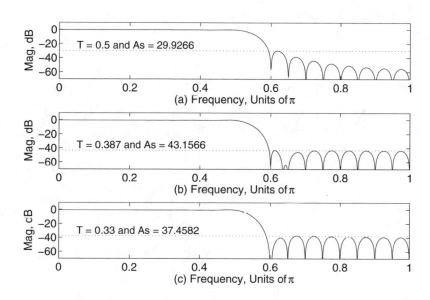

Figure 2.44: (a) A length-40 FIR having $\omega_p = 0.5\pi$, $\omega_s = 0.6\pi$, designed by the frequency sampling method, with one transition coefficient T valued at 0.5; (b) Same as (a), but with the ideal-valued T of 0.387; (c) same as (a), but with $T = 0.33$. R_P was 0.67 dB for $T = 0.387$.

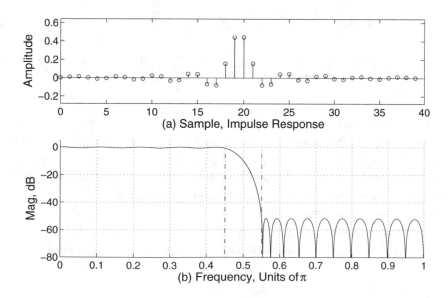

Figure 2.45: A length-40 equiripple FIR having $\omega_p = 0.5\pi$, $\omega_s = 0.6\pi$, $R_P = 0.69$ dB, and $A_S = 50.95$ dB.

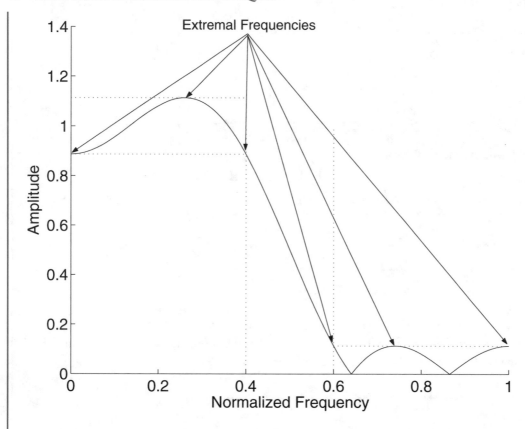

Figure 2.46: An equiripple lowpass filter having wp = 0.4 and ws = 0.6. Extremal frequencies are located at normalized frequencies 0, 0.2625, 0.4, 0.6, 0.7375, 1.0.

2.7.3   ALTERNATION THEOREM

An important theorem regarding the frequency response of an FIR is the Alternation Theorem, which states essentially that a set of filter coefficients exist which will result in a frequency response having equal ripple amplitudes, and that in fact the deviation amplitude for all ripples is limited to some value δ, with the sign of δ alternating between adjacent extremal frequencies.

Formal Statement of Alternation Theorem

If S is any closed subset of the closed frequency interval $[0, \pi]$, $P(\omega)$ is the unique minimax approximation to $H_{dr}(\omega)$ over S if $E(\omega)$ exhibits at least $(R + 2)$ unique extremal frequencies $[\omega_0, \omega_1, ...\omega_{R+1}]$ and

$$E(\omega_{i+1}) = -E(\omega_i) = \pm\delta = \pm \max_{S} |E(\omega)|$$

with

$$\omega_0 < \omega_1 ... < \omega_{R+1} \in S$$

2.7.4 A COMMON DESIGN PROBLEM FOR ALL LINEAR PHASE FILTERS

In order to simplify the design algorithm, it is desirable to have a common form for the real frequency response of the four different types of Linear Phase FIRs, which have real frequency responses $H_r(\omega)$ as described below.

Type I:

$$H_r(\omega) = \sum_{k=0}^{(L-1)/2} \alpha[k] \cos(\omega k)$$

with

$$\alpha[k] = \begin{cases} h(M) & k = 0 \\ 2h(M-k) & k = 1, 2...M \end{cases}$$

where $M = (L-1)/2$ and L is the filter length.

Type II:

$$H_r(\omega) = \sum_{k=1}^{L/2} b[k] \cos(\omega(k-0.5)) \tag{2.11}$$

with

$$b[k] = 2h(L/2 - k)$$

where $k = 1, 2...L/2$. Eq. (2.11) can further be rewritten as

$$H_r(\omega) = \cos(\omega/2) \sum_{k=1}^{L/2} b'[k] \cos(\omega k) \tag{2.12}$$

where

$$\begin{aligned} b'[0] &= b[1]/2 \\ b'[L/2 - 1] &= 2b[L/2] \end{aligned}$$

and

$$b'[k] = 2b[k] - b'[k-1]; \quad k = 1, 2, ...(L/2) - 2$$

Type III:

$$H_r(\omega) = \sum_{k=1}^{M} c[k] \sin(\omega\, k) \tag{2.13}$$

where

$$c[k] = 2h[M - k]; \quad k = 1, 2, ...M$$

and $M = (L - 1)/2$. The above can be rewritten as

$$H_r(\omega) = \sin(\omega) \sum_{k=0}^{(L-3)/2} c'[k] \cos(\omega k) \tag{2.14}$$

where $c'[n]$ is linearly related to $c[n]$ (the relationship may be found in [5]).

Type IV:

$$H_r(\omega) = \sum_{k=1}^{L/2} d[k] \sin(\omega(k - 0.5)) \tag{2.15}$$

where

$$d[k] = 2h[L/2 - k]; \quad k = 1, 2, ...L/2$$

The above can be rewritten as

$$H_r(\omega) = \sin(\omega/2) \sum_{k=0}^{L/2-1} d'[k] \cos(\omega k) \tag{2.16}$$

where $d'[n]$ is linearly related to $d[n]$ (the relationship may be found in [5]).

The above real frequency responses can be rewritten as the product of two functions $Q(\omega)$ and $P(\omega)$ according to the following table:

Type	$Q(\omega)$	$P(\omega)$	K	
I	1	$\sum_0^K \alpha[n]\cos(\omega n)$	$\frac{L-1}{2}$	
II	$\cos(\omega/2)$	$\sum_0^K b'[n]\cos(\omega n)$	$\frac{L}{2} - 1$	(2.17)
III	$\sin(\omega)$	$\sum_0^K c'[n]\cos(\omega n)$	$\frac{L-3}{2}$	
IV	$\sin(\omega/2)$	$\sum_0^K d'[n]\cos(\omega n)$	$\frac{L}{2} - 1$	

2.7.5   WEIGHTED ERROR FUNCTION

It is possible to obtain a filter that has different ripple magnitudes in the pass- and stop- bands. If the maximum magnitude of passband ripple is δ_1 and the maximum magnitude of stopband ripple is δ_2, then the needed weight vector $W(\omega)$ is defined as

$$W(\omega) = \begin{cases} \delta_2/\delta_1 & \omega \text{ in passband} \\ 1 & \omega \text{ in stopband} \end{cases}$$

The weighted approximation error $E(\omega)$ is then defined as

$$E(\omega) = W(\omega)[H_{dr}(\omega) - H_r(\omega)]$$

which becomes

$$E(\omega) = W(\omega)[H_{dr}(\omega) - Q(\omega)P(\omega)]$$

This expression can be modified to

$$E(\omega) = W(\omega)Q(\omega)[H_{dr}(\omega)/Q(\omega) - P(\omega)]$$

and, setting

$$\hat{W}(\omega) = W(\omega)Q(\omega)$$

and

$$\hat{H}_{dr}(\omega) = \frac{H_{dr}(\omega)}{Q(\omega)}$$

we obtain as the net weighted error expression

$$E(\omega) = \hat{W}(\omega)[\hat{H}_{dr}(\omega) - P(\omega)]$$

When the proper extremal frequencies ω_n are known for a certain equiripple FIR design, it is true that

$$E(\omega_n) = \hat{W}(\omega_n)[\hat{H}_{dr}(\omega_n) - P(\omega_n)] = (-1)^n \delta$$

for $n = 0:1:K +1$ where K is as given in Table (2.17). The set of $K + 2$ equations can be rewritten as

$$P(\omega_n) + (-1)^n \delta/\hat{W}(\omega_n) = \hat{H}_{dr}(\omega_n)$$

Replacing $P(\omega_n)$ with the equivalent cosine summation expression, we get

$$\sum_{k=0}^{K} \alpha_k \cos(\omega_n k) + \frac{(-1)^n \delta}{\hat{W}(\omega_n)} = \hat{H}_{dr}(\omega_n)$$

where $n = 0{:}1{:}K+1$, and K is as given in Table (2.17).

In expanded matrix form this would appear as

$$
\begin{bmatrix}
1 & \cos(\omega_0) & \cdots & \cos(K\omega_0) & \frac{1}{\hat{W}(\omega_0)} \\
1 & \cos(\omega_1) & \cdots & \cos(K\omega_1) & \frac{-1}{\hat{W}(\omega_1)} \\
\vdots & \vdots & \vdots & \vdots & \vdots \\
1 & \cos(\omega_{K+1}) & \cdots & \cos(K\omega_{K+1}) & \frac{(-1)^{K+1}}{\hat{W}(\omega_{K+1})}
\end{bmatrix}
\begin{bmatrix}
\alpha[0] \\
\alpha[1] \\
\vdots \\
\alpha[K] \\
\delta
\end{bmatrix}
=
\begin{bmatrix}
\hat{H}_{dr}(\omega_0) \\
\hat{H}_{dr}(\omega_1) \\
\vdots \\
\hat{H}_{dr}(\omega_{K+1})
\end{bmatrix}
\tag{2.18}
$$

Written in more compact form this would be

$$[WP][A] = [H]$$

where WP is the cosine-and-reciprocal-weights matrix, A is the vector of α values and δ, and H is the $\hat{H}_{dr}$ vector.

2.7.6 REMEZ EXCHANGE ALGORITHM

Unfortunately, the values for α, δ, and ω_n in matrix equation (2.18) are unknown. However, by making a guess or assumption for the initial values of the ω_n, it is possible to solve for the vector comprising the α values and δ. When this is done, the frequency response can be computed using the values of α, and a search of the frequency response is conducted for the next estimate of the extremal values. This procedure is known as the Remez Exchange algorithm, discussed and demonstrated immediately below. There are several ways to solve for the values of α and δ. One direct, but computationally expensive way is to compute the inverse (actually, the pseudo-inverse) of matrix WP:

$$[WP]^{-1}[WP][A] = [WP]^{-1}[H]$$

which reduces to

$$[A] = [WP]^{-1}[H]$$

The well-known Parks-McClellan algorithm analytically computes the value for δ, generates a set of extremal points using δ and the assumed or current-estimated extremal frequencies, interpolates on a fine grid between the extremal points to produce a frequency response, computes the error function, and then conducts a search of the error function for the next estimates of the extremal frequencies. The manner of interpolating between the extremal points is known as Barycentric Lagrangian interpolation. We will discuss the MathScript implementation of the Parks-McClellan algorithm below, after further exploration of the basic Remez Exchange concept.

Example 2.25. Write a program that will receive a set of band limits for a lowpass filter, a filter length, and a set of extremal frequencies, which will solve for delta and the alpha coefficients, and display the corresponding $P(\omega)$ in a way that allows visual selection of the next set of extremal frequencies.

The following program has been written for one specific filter design; the exercises at the end of the chapter will direct the reader in various projects to expand and generalize the code. In order to keep things conceptually simple, the values of α and δ are determined using the matrix equation method, which is acceptable for shorter filter lengths and modern, relatively fast computers.

```
function LVManualLPFRemezExch(L,LenGrid,wp,ws,curXFrqs)
% LVManualLPFRemezExch(9,145,0.45,0.55,[0,0.225,0.45,...
0.55,0.775,1])
NormFrGrid = [0:1:LenGrid-1]/(LenGrid-1); wc = (wp+ws)/2;
FrGrid = pi*NormFrGrid; WtVec = ones(1,LenGrid);
XFrindOnFG = round(curXFrqs*(LenGrid-1)+1);
Hdr = ones(1,round(wc*(LenGrid-1)));
Hdr = [Hdr, zeros(1,LenGrid-length(Hdr))];
Q = 1; kLim = (L-1)/2; WMat(1:kLim+2,1) = 1; Hdr = Hdr./Q;
WtVec = WtVec.*Q; k = 0:1:kLim+1;
Num = ((-1).^k); Denom = WtVec(1,XFrindOnFG);
WMat(k+1,2:kLim+1) = cos(pi*curXFrqs(1,k+1)'*([1:kLim]));
WMat(k+1,kLim+2) = (Num./Denom)';
HdrVec = Hdr(round((LenGrid-1)*[curXFrqs]+1));
AlDelVec = pinv(WMat)*(HdrVec)';
delta = AlDelVec(length(AlDelVec)),
P = 0; for pCtr = 1:length(AlDelVec)-1;
P = P + AlDelVec(pCtr)*(cos(FrGrid*(pCtr-1)) ); end;
E = WtVec.*([Hdr - P]); figure(8); clf;
subplot(211); hold on; ad = abs(delta);
loBndX = [0:1:round((LenGrid-1)*wp)]/(LenGrid-1);
hiBndX = [round((LenGrid-1)*ws):1:LenGrid-1]/(LenGrid-1);
plot(loBndX,E(1,1:round(wp*(LenGrid-1))+1));
line([0,max(loBndX)],[ad, ad]); line([0,max(loBndX)],[-ad, -ad]);
xlabel('Norm Freq'); ylabel('Err (Passband)');
subplot(212); hold on;
plot(hiBndX,E(1,round(ws*(LenGrid-1))+1:LenGrid));
line([ws,1],[ad, ad]); line([ws,1],[-ad, -ad]);
xlabel('Norm Freq'); ylabel('Err (Stopband)');
Hr = Q.*P; figure(9); LenHr = length(Hr);
plot([0:1:LenHr-1]/LenHr,Hr);
xlabel('Norm Freq'); ylabel('Amp')
```

The code above has for *curXFrqs* (the current set of extremal frequencies) an initial guess of linearly spaced normalized frequencies that include the two transition band edges as well as 0 and 1.0; the remaining two needed values of a total of six needed values ((9-1)/2 + 2 = 6) were spaced equally within the pass- and stop- bands.

A script which can receive several arguments to perform the same functions as the code above for Type I and II lowpass and highpass filter types is

$$LVDemoRemez(wp, ws, As, Rp, L, PassbandType, NormXFr, EqWt)$$

To demonstrate the principles of the Remez Exchange algorithm, we can start with the following call, which uses extremal frequencies that are linearly-spaced in the pass- and stop- bands for the initial guess:

LVDemoRemez(0.45,0.55,50,0.2,9,1,[0,0.225,0.45,0.55,0.775,1],1)

the result of which is shown in Fig. 2.47.

The initial guess of extremals generally includes the normalized frequencies of 0 and 1, and all transition band edges. In subsequent extremal searches, transition band edges are usually included since the design expects extremals at the transition band edges in the final solution. Extremals at normalized frequencies 0 and 1 may or may not be present after the initial guess. To be systematic and accurate, an extremal candidate list should be made which contains at least the frequency and magnitude of each candidate extremal. Another useful parameter is whether each candidate extremal is a local maximum ("positive") or minimum ("negative"), since a proper set of extremals must alternate between being positive and negative. When the list is complete, the candidates having the highest ($kLim$ + 2) magnitudes (and which properly alternate in sign) are chosen as the next set of extremals. A detailed discussion of extremal frequency choice can be found in [6].

Continuing with the present example, for our next call, we use extremal frequencies estimated by inspecting plots (a) and (b) in Fig. 2.47. These appear to be 0, 0.27, 0.45, 0.55, 0.675, and 1.0. We thus make the call

LVDemoRemez(0.45,0.55,50,0.2,9,1,[0,0.27,0.45,0.55,0.675,1],1)

which results in Fig. 2.48, which shows all extremals nearly within the bounds of $\pm\delta$; one or two more iterations with very small changes should bring the design to complete convergence.

MathScript has the functions *remez* and *firpm* (which may be used interchangeably), both of which implement the Parks-McClellan algorithm. The basic format is as follows:

$$b = firpm(N, F, A, W)$$

where b is the desired impulse response, N is the order (one less than the desired FIR length), F is a vector of band edges in normalized frequencies (from 0 to 1), A is a vector (half the length of F) that gives desired response amplitude for each band (i.e., passband and stopband) defined by the

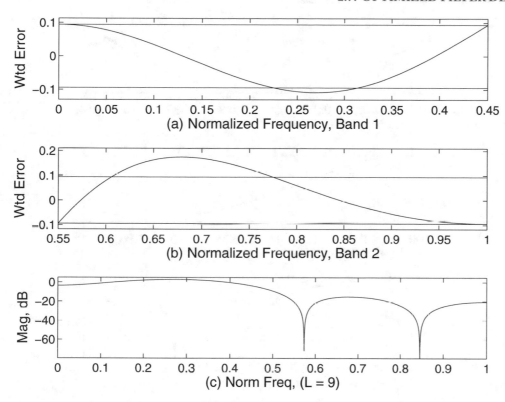

Figure 2.47: (a) Error signal in passband of a length-9 lowpass FIR; (b) Error signal in stopband of same; (c) Frequency response of filter based on the extremals [0,0.225,0.45,0.55,0.775,1]; Note the new extremal frequencies (estimated by visual inspection of the plots) at 0, 0.27, 0.45, 0.55, 0.675, and 1.0.

pairs of band edges given in vector F. W is a weight vector telling what weights to give the error in each band; W has the same length as A. The following are a few example calls:

lowpass	**b = firpm(18,[0,0.5,0.6,1],[1,1,0,0])**
highpass	**b = firpm(18,[0,0.5,0.6,1],[0,0,1,1])**
bandpass	**b = firpm(18,[0,0.35,0.4,0.55,0.6,1],[0,0,1,1,0,0])**
bandstop	**b = firpm(18,[0,0.35,0.4,0.55,0.6,1],[1,1,0,0,1,1])**

The approximate necessary filter length (one greater than the filter order, which is specified in the above calls) to meet certain desired values of A_s and R_p, can be computed according to a formula by J. F. Kaiser, which is

$$L = \frac{-20\log_{10}(\sqrt{\delta_1\delta_2}) - 13}{14.6\Delta f} \tag{2.19}$$

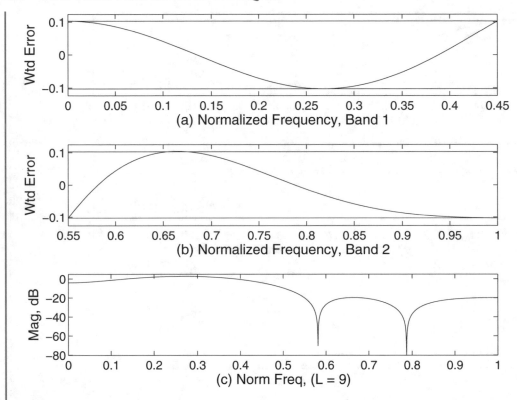

Figure 2.48: (a) Error signal in passband of a length-9 lowpass FIR; (b) Error signal in stopband of same; (c) Frequency response of filter based on the extremals [0,0.27,0.45,0.55,0.675,1], which shows the weighted error signal nearly within the limits of $\pm\delta$; one or two more iterations with minor changes in extremal estimates should be adequate to achieve complete convergence.

where $\Delta f = (\omega_p - \omega_s)/2\pi$, where ω_p and ω_s are radian frequencies, i.e., normalized frequencies multiplied by π.

To design a Hilbert transformer or a differentiator, the input argument *ftype* is necessary. Here are several sample calls:

b = remez(18,[0.1,0.9],[1,1],'Hilbert')
b = remez(17,[0:(1/7):1],[0:(1/7):1],'differentiator')

The value for L given by Eq. (2.19) is only approximate, and the needed value is often somewhat greater. To design a filter to meet certain specifications, it is necessary to run the design several times with increasing values of L, testing the result after each design run. This can be done programmatically as shown in the following example:

Example 2.26. Write code that will design an equiripple lowpass filter that meets certain specifications of A_s, R_p, ω_p, and ω_s.

```
[actRp,actAs,WF] = LVDesignEquirippLPF(Rp,As,wp,ws)
% [actRp,actAs,WF] = LVDesignEquirippLPF(0.2,60,0.45,0.55)
Rfac= 10^(-Rp/20); DeltaP = (1-Rfac)/(1+Rfac);
DeltaS = (1+DeltaP)*10^(-As/20);
deltaF = abs(ws - wp)/2;
compL = (-20*log10(sqrt(DeltaP*DeltaS)) - ...
13)/(14.6*deltaF) +1;
L = round(compL), Ord = L - 2; LenFFT = 2^15;
SBAtten = 0; PBR = 100;
while (SBAtten < As)|(PBR > Rp); Ord = Ord + 1
WF = firpm(Ord,[0,wp,ws,1],[1,1,0,0],[DeltaS/DeltaP,1]);
fr = abs(fft(WF,LenFFT));
fr = fr(1,1:LenFFT/2+1)/(max(abs(fr)));
Lfr = length(fr); PB = fr(1,1:round(wp*(Lfr-1)));
SB = fr(1,round(ws*(Lfr-1))+1:Lfr);
PBR = -20*log10(min(PB) + eps)
SBAtten = -20*log10(max(SB) + eps)
end; Fin_L = Ord + 1, actRp = PBR;
actAs = SBAtten; figure(77);
plot([0:1:LenFFT/2]/(LenFFT/2), 20*log10(fr+eps));
xlabel('Normalized Frequency (Units of \pi)')
ylabel('Magnitude, dB'), axis([0,1,-(As+20),5])
```

The result from running the above is shown in Fig. 2.49.

Example 2.27. Design a bandpass filter having band edges at [0, 0.4, 0.45, 0.65, 0.7, 1] with corresponding amplitudes of [0, 0, 1, 1, 0, 0], $As = 60$ dB, and $Rp = 0.2$ dB.

The following code is similar to that of the previous example except that the stopband evaluation is done for both stopbands.

```
function LVDesignEquirippBPF(Rp,As,ws1,wp1,wp2,ws2)
% LVDesignEquirippBPF(0.2,60,0.4,0.45,0.65,0.7)
BndLm = [0,ws1,wp1,wp2,ws2,1];
Rfac= 10^(-Rp/20); DeltaP = (1-Rfac)/(1+Rfac);
DeltaS = (1+DeltaP)*10^(-As/20);
deltaF = abs(wp1 - ws1)/2;
compL = (-20*log10(sqrt(DeltaP*DeltaS)) - ...
```

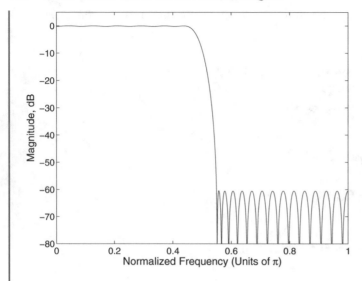

Figure 2.49: A lowpass filter designed to have the minimum length needed (53) to meet an A_s of 60 dB, R_p = 0.2 dB, with ω_p = 0.45 and ω_s = 0.55.

```
13)/(14.6*deltaF) +1;
L = round(compL), Ord = L - 2; LenFFT = 2^15;
SBAtten = 0; PBR = 100;
while (SBAtten < As)|(PBR > Rp); Ord = Ord + 1;
WF = firpm(Ord,BndLm,[0,0,1,1,0,0],[1,DeltaS/DeltaP,1]);
fr = abs(fft(WF,LenFFT));
fr = fr(1,1:LenFFT/2+1)/max(fr);
Lfr = length(fr);
PB = fr(1,round(wp1*Lfr):round(wp2*Lfr));
PBR = -20*log10(min(PB)+eps),
SB1 = fr(1,1:round(ws1*Lfr));
SB2 = fr(1,round(ws2*Lfr):Lfr);
SBAtten1 = -20*log10(max(SB1)+eps);
SBAtten2 = -20*log10(max(SB2)+eps);
SBAtten = min([SBAtten1,SBAtten2]),
end; Fin_L = Ord + 1, Fin_Rp = PBR,
Fin_As = SBAtten, figure(8); clf;
plot([0:1:LenFFT/2]/(LenFFT/2), 20*log10(fr+eps));
xlabel('Normalized Frequency (Units of \pi)')
ylabel('Magnitude, dB'); axis([0,1,-(As+20),5])
```

The result from running the above is shown in Figure 2.50.

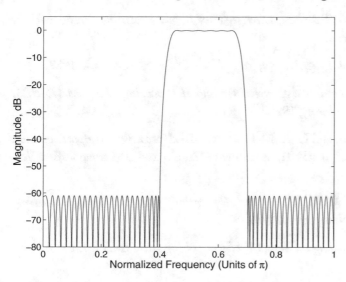

Figure 2.50: A bandpass filter designed to have the minimum length needed (110) to meet an A_s of 60 dB, R_p = 0.2 dB, with ω_{s1} = 0.4, ω_{p1} = 0.45, ω_{p2} = 0.65, and ω_{s2} = 0.7.

Example 2.28.   Design a Hilbert Transformer having its passband edges at 0.1 and 0.9, with Rp = 0.2 dB

The following code is straightforward, incrementing the order by two (to maintain symmetry) until the passband ripple is below the value specified for R_P.

```
function LVDesignEquirippHilbert(Rp,wp1,wp2)
% LVDesignEquirippHilbert(0.2,0.1,0.9)
PBR = 100; Ord = 10; LenFFT = 2^13;
figure(99); while (PBR > Rp); Ord = Ord + 2;
b = remez(Ord,[0.1,0.9],[1,1],'Hilbert');
y = abs(fft(b,LenFFT));
y = y(1,1:LenFFT/2+1)/(max(abs(y)));
LenGrid = LenFFT/2;
PB = y(1,round(wp1*LenGrid)+1:round(wp2*LenGrid)+1);
PBR = -20*log10(min(PB)+eps);
plot([0:1:LenGrid]/LenGrid,20*log10(y+eps))
xlabel(['Normalized Frequency (Units of \pi)'])
ylabel('Magnitude, dB'); L = Ord + 1,
axis([0,1,-40,5]); pause(0.3); end
Final_Rp = PBR, Final_L = Ord + 1
```

2.8   REFERENCES

[1] Alan V. Oppenheim and Ronald W. Schaefer, *Discrete-Time Signal Processing*, Prentice-Hall, Englewood Cliffs, New Jersey 07632, 1989.

[2] T. W. Parks and C. S. Burrus, *Digital Filter Design*, John Wiley & Sons, New York, 1987.

[3] James H. McClellan et al, *Computer-Based Exercises for Signal Processing Using MATLAB5*, Prentice-Hall, Upper Saddle River, New Jersey, 1998.

[4] Lawrence R. Rabiner, Bernard Gold, and C. A. McGonegal, *"An Approach to the Approximation Problem for Nonrecursive Digital Filters,"* IEEE Transactions on Audio and Electroacoustics, Vol. AU-18, pp. 83-106, June 1970.

[5] Lawrence R. Rabiner and Bernard Gold, *Theory and Application of Digital Signal Processing*, Prentice-Hall, Inc. Englewood Cliffs, New Jersey, 1975.

[6] Vinay K. Ingle and John G. Proakis, *Digital Signal Processing Using MATLAB V.4*, PWS Publishing Company, Boston, 1997.

[7] John G. Proakis and Demitris G. Manolakis, *Digital Signal Processing, Principles, Algorithms, and Applications (Third Edition)*, Prentice Hall, Upper Saddle River, New Jersey 07458, 1996.

[8] Andreas Antoniou, *Digital Filters, Analysis, Design, and Applications (Second Edition)*, McGraw-Hill, New York, 1993.

2.9   EXERCISES

1. Devise a script conforming to the following call syntax:

$$LVxFRviaCirConDFTs(LL, LenWin, wintype)$$

where LL is the length of the simulated ideal lowpass filter impulse response, $LenWin$ is the length of the window used to truncate the ideal lowpass filter impulse response, and $wintype$ determines the type of window used to truncate the ideal filter: pass $wintype$ as 1 for rectangular, 2 for Hamming, 3 for Blackman, and 4 for Kaiser with $\beta = 10$. The script should create an ideal lowpass filter impulse response of length $LL = L$, obtain its DFT, create a window of the same length consisting of zeros except for the central samples, which comprise the window selected by $wintype$ having length $LenWin$, obtain its DFT, and obtain the net truncated filter response as the circular convolution of the DFTs of the ideal lowpass filter and the window. A figure having the same plots shown, for example, in any of Figs. 2.6, 2.7, and 2.8, plots (a), (b), and (d), should be created.

Once the script has been written and tested, pick a value of LL of about 4096, $wintype=$ 1, and run the script for values of $LenWin=10, 50, 200,$ and 500. Compare results, noting the transition width for each case and the minimum stopband attenuation. Repeat this for the Hamming, Blackman, and Kaiser(10) window by specifying $wintype$ as 2, 3, and 4, respectively.

Test Calls:

(a) **LVxFRviaCirConDFTs(4096,15,1)**

(b) **LVxFRviaCirConDFTs(4096,50,1)**

(c) **LVxFRviaCirConDFTs(4096,200,1)**

(d) **LVxFRviaCirConDFTs(4096,500,1)**

2. Create and test a script conforming to the call syntax below, and as described in the text, which outlines a procedure for creating the script; the reader may use the suggested procedure or devise one functionally equivalent.

function LVxFIRViaWinIdealLPF(FiltType, BandEdgeVec, ...
typeWin, As, Rp)
% FiltType: 1 = Lowpass, 2 = Highpass, 3 = Bandpass, 4 = Notch
% BandEdgeVec must have 2 or 4 values; it must have length 2
% when defining ws and wp for an LPF or an HPF; it must have
% length 4 for a BPF or Notch filter. typeWin is one of the
% following functions in string format, i.e., surrounded by
% single quotes such as 'rectwin', 'kaiser', 'blackman',
% 'hanning', 'hamming', or 'bartlett'
% As is minimum desired stopband attenuation in dB, as a
% positive number such as 51, 67, etc.
% Test calls:
% Lowpass filters
% LVxFIRViaWinIdealLPF(1, [0.2,0.3], 'kaiser', 80, 0.1)
% LVxFIRViaWinIdealLPF(1, [0.2,0.3], 'boxcar', 21, 0.1)
% LVxFIRViaWinIdealLPF(1, [0.2,0.3], 'hanning', 44, 0.1)
% LVxFIRViaWinIdealLPF(1, [0.2,0.3], 'hamming', 53, 0.1)
% LVxFIRViaWinIdealLPF(1, [0.2,0.3], 'blackman', 74, 0.1)
% Highpass filters
% LVxFIRViaWinIdealLPF(2, [0.5,0.6], 'kaiser', 40, 0.1)
% LVxFIRViaWinIdealLPF(2, [0.5,0.6], 'kaiser', 60, 0.1)
% LVxFIRViaWinIdealLPF(2, [0.5,0.6], 'kaiser', 80, 0.1)
% Bandpass filters
% LVxFIRViaWinIdealLPF(3, [0.2,0.3,0.5,0.6], 'kaiser', 40, 0.1)
% LVxFIRViaWinIdealLPF(3, [0.2,0.3,0.5,0.6], 'kaiser', 60, 0.1)
% LVxFIRViaWinIdealLPF(3, [0.2,0.3,0.5,0.6], 'kaiser', 80, 0.1)
% Bandstop filters
% LVxFIRViaWinIdealLPF(4, [0.2,0.3,0.5,0.6], 'kaiser', 40, 0.1)
% LVxFIRViaWinIdealLPF(4, [0.2,0.3,0.5,0.6], 'kaiser', 60, 0.1)
% LVxFIRViaWinIdealLPF(4, [0.2,0.3,0.5,0.6], 'kaiser', 80, 0.1)

Here is a possible design procedure for the script *LVxFIRViaWinIdealLPF*:

a) Write a function *LVxTrunIdealLowpass* that receives a cutoff frequency ω_c in radians and a desired length L and returns to a calling function a truncated ideal lowpass filter impulse response of the given length L and cutoff frequency ω_c.

b) Write a function *LVxLPFViaWindowedSincND* that receives as arguments values for ω_p, ω_s, L, and a desired window type to use, *typeWin* (including the parameter β if *typeWin* is Kaiser). Using these parameters, the script must then obtain a truncated ideal lowpass filter impulse response of length L by calling the function *LVxTrunIdealLowpass*, apply the chosen window, evaluate the frequency response of the filter, and compute and return to a calling function the impulse response, and the actual passband and stopband ripple values.

c) Write a function *LVxDesignLPFViaWindowedSincND* that receives the arguments ω_p, ω_s, *typeWin*, and the desired A_s, estimates an initial value for L (and computes β if *typeWin* is Kaiser), and repeatedly calls *LVxLPFViaWindowedSincND*, gradually increasing L to the lowest value that can produce the desired value of A_s. This script can be used directly to design a windowed lowpass filter. Other filter types (highpass, bandpass, bandstop) must be derived using various combinations of windowed lowpass filters, which is performed by the next listed script, *LVxFIRViaWinIdealLPF*.

d) Write a script *LVxFIRViaWinIdealLPF* that receives arguments consisting of the desired filter type (lowpass, highpass, bandpass, bandstop), a vector of band edges of length two for lowpass and highpass filters, and length four for bandpass and bandstop filters; a desired value of A_s, and a desired window type. A desired maximum passband ripple may also be specified, and the script will return the actual value of passband ripple for comparison to the design goal.

3. Write the m-code for the script *LVxFilterViaCosineFormula*, as described and illustrated in the text, and test it with the given sample calls.

```
function WF = LVxFilterViaCosineFormula(Type,Bin0,PosBins)
% WF is the output impulse response designed by the script
% A Type I filter is obtained by passing Type as 1
% A Type II filter is obtained by passing Type as 2
% Bin0 is the sample amplitude for frequency zero.
% PosBins are sample amplitudes for frequencies 1 to (L-1)/2 for
% odd length, or 1 to L/2-1 for even length.
% Sample calls
% WF = LVxFilterViaCosineFormula(2,[1],[1,1,1,1,0,0,0,0])
% WF = LVxFilterViaCosineFormula(2,[0],[0,0,0,1,1,1,0,0])
% WF = LVxFilterViaCosineFormula(2,[0],[0,0,0,0,1,1,1,1,1])
```

4. Write the m-code for the script *LVxFIRViaWholeSines* as illustrated and described in the text and below, and test it using the sample calls given. Sine summation formulas for Type-II and Type-IV filters can be found in the Appendices (use the formulas for whole sine correlators rather than half-sine correlators)

```
function WF = LVxFIRViaWholeSines(AkPos,AkLOver2,L)
% Returns an impulse response WF for a Type III or Type IV
```

```
% linear phase FIR.
% AkPos are sample amplitudes for frequencies 1 to (L-1)/2 for
% odd length, or 1 to L/2-1 for even length. AkLOver2 is passed
% as 0 or the empty matrix [] for odd values of L, and as a desired
% amplitude for even values of L, which is the desired filter length
% Sample calls:
% WF = LVxFIRViaWholeSines([ones(1,38)],[],77); % Hilb
% WF = LVxFIRViaWholeSines([0.8,ones(1,36),0.55],[],77); transition
% values approximately optimized to reduce ripple.
% Type-IV differentiators
% WF = LVxFIRViaWholeSines([1:1:11]*pi/12,[12]*pi/12,24);
% WF = LVxFIRViaWholeSines([1:1:38]*pi/39,[39]*pi/39,78);
% Type-III differentiator
% WF = LVxFIRViaWholeSines([1:1:38]*pi/39,[],77);
```

5. Convert the script/function given in the text

$$LVxDifferentiatorLen24$$

into a generic script/function to generate the impulse response for a Type-IV differentiator of arbitrary even length, having the following call syntax:

$$ImpResp = LVxDifferentiatorTypeIV(L)$$

Your function should compute and display the impulse response ($ImpResp$) and corresponding frequency response for any given even filter length L.

6. Write and test a script that will create a triangle wave as a test signal, and convolve it with a differentiator, which the script creates, to generate a square wave. The script should conform to the following call syntax:

$$LVxTriang2SquareViaDiff(SR, HiHrm, Fo, Ldiff)$$

where SR is a sample rate that determines how many samples the test signal will have, $HiHrm$ (an odd integer) is the highest harmonic to add in to synthesize the test waveform having a fundamental frequency of Fo, according to the formula for generating a triangle waveform, which is

$$WF = \sum_{n=1}^{(HiHrm+1)/2} (1/(2n-1)^2)\cos(2\pi n F_o t)$$

where $t = [0:1:SR]/SR$. $Ldiff$ is the length of the differentiator which is to be convolved with the test signal WF to yield a squarewave.

The call

LVxTriang2SquareViaDiff(4000,181,20,94)

should result in substantially the same plot as shown in Fig. 2.43.

7. Write the m-code for the following function, which is described and illustrated in the text with syntax below:

> function LVxHilbertViaConvolution(TestSeqLen,TestWaveType,...
> FilterLen,TestSigFreq,FDMethod,UseWin)
> % TestSeqLength is the desired test sequence length;
> % TestWaveType = '1' for sawtooth, and '2' for squarewave;
> % FilterLen = length of the Hilbert trans. imp resp made directly
> % from time domain formula;
> % TestSigFreq = fundam. freq of (truncated) sawtooth or square
> % wave used as the test signal;
> % FDMethod = 1 for an All-Real FD Mask, or 2 for All-Imaginary
> % mask the same length as.the test signal, and is converted using
> % the ifft to a TD Hilbert transformer which is convolved with the
> % test signal.
> % UseWin = 0 to use the raw TD Hilbert impulse response, or 1 to
> % window it with a Kaiser window, Beta = 5
> % Test call:
> % LVxHilbertViaConvolution(64, 1, 19, 5, 2,0)

8. Write a script that designs an equiripple notch filter to meet certain specifications as given in the following call, and test it with the given calls, noting, for each call, the length of the resulting filter.

> function [R,A,b] = LVxDesignEquirippNotch(Rp,As,...
> wp1,ws1,ws2,wp2)
> % Rp is the maximum design passband ripple in positive dB
> % and As is the minimum design stopband attenuation in dB,
> % wp1, ws1,ws2, and wp2 specify the 1st passband, 1st
> % stopband, 2nd stopband, & 2nd passband frequencies,
> % respectively. R and A are the realized values of Rp
> % and As, respectively, and b is the vector of designed filter
> % coefficients.
> % Test calls:
> % [R,A,b] = LVxDesignEquirippNotch(0.2,40,0.2,0.3,0.5,0.6)
> % [R,A,b] =LVxDesignEquirippNotch(0.2,55,0.2,0.3,0.5,0.6)
> % [R,A,b] =LVxDesignEquirippNotch(0.2,70,0.2,0.3,0.5,0.6)
> % [R,A,b] =LVxDesignEquirippNotch(0.2,40,0.2,0.25,0.4,0.45)
> % [R,A,b] =LVxDesignEquirippNotch(0.2,55,0.2,0.25,0.4,0.45)

% **[R,A,b] =LVxDesignEquirippNotch(0.2,70,0.2,0.25,0.4,0.45)**

9. For each of the equiripple notch filters designed in the previous exercise, design a minimum-length notch filter meeting the same specifications using the Windowed-Ideal LPF technique with a Kaiser window. Which technique results in the shorter filter in general? Compare the filter lengths of the equiripple and Kaiser filters for the A_S = 40 dB case, the A_S = 55 dB case, and the A_S = 70 dB case, and note the relative (or percentage) length difference between the Kaiser and equiripple designs, then compare the two lengths for the A_S = 55 dB and A_S = 70 dB cases and note again the relative or percentage length differences. You should find that at higher values of A_S, the efficiency of the equiripple design compared to the Kaiser windowed design is greater than for smaller values of A_S.

10. Using the script *LVDemoRemez*, for each given call, iteratively manually estimate successive sets of extremals, attempting to achieve convergence, i.e., an equiripple filter.

(a) **LVDemoRemez(0.45,0.55,40,0.5,33,1,[],1)**
(b) **LVDemoRemez(0.45,0.55,40,0.5,9,1,[0,0.2875,0.45,0.55,...**
0.7125,1],1)
(c) **LVDemoRemez(0.65,0.75,40,0.5,19,1,[0,0.135,0.265,0.385,...**
0.5,0.6,0.65,0.75,0.8,0.895,1],1)
(d) **LVDemoRemez(0.65,0.75,40,0.5,20,1,[0,0.106,0.208,0.315,...**
0.41,0.519,0.6062,0.65,0.75,0.8,0.9063],1)
(e) **LVDemoRemez(0.75,0.65,60,0.5,19,2,[0,0.125,0.27,0.385,...**
0.5,0.6,0.65,0.75,0.795,0.892,1],1)
(f) **LVDemoRemez(0.75,0.65,40,0.5,19,2,[0,0.1,0.22,0.35,0.46,...**
0.56,0.61,0.75,0.79,0.88,1],0)
(g) **LVDemoRemez(0.75,0.65,40,0.5,19,2,[0,0.138,0.269,...**
0.3875,0.5,0.6,0.65,0.75,0.8,0.894,1],1)

11. Attempt to achieve the three optimum transition band values $T1$, $T2$, and $T3$ in either of the LabVIEW VIs

DemoHPFOptimizeXitionBandsVI

DemoHPFOptimizeXitionBandsPrecVI

Use the following procedure: Set $T1$, $T2$, and $T3$ to 1.0, then gradually reduce $T1$ until the maximum stopband attenuation is achieved. Then reduce $T2$ by 0.05, and adjust $T1$ until a new minimum is achieved. Keep reducing $T2$ in steps of 0.05 and reoptimizing $T1$ until the best stopband attenuation has been achieved using only $T1$ and $T2$. Then reduce $T3$ by 0.05, then perform again the entire $T2$, $T1$ optimization procedure. Repeat the entire general procedure until the best stopband attenuation has been achieved. In an actual computer search algorithm, of course, the step size would need to be much smaller than 0.05, but the procedure outlined above illustrates one way to determine approximate optimum values.

12. Write a script that designs an equiripple highpass filter according to the following function description:

> **function [actRp,actAs,WF] = LVxDesignEquirippHPF(Rp,As,...**
> **ws,wp)**
> **% Rp and As are the maximum passband ripple and minimum**
> **% stopband attenuation, respectively.**
> **% ws and wp are the stopband and passband edges, specified**
> **% in normalized frequency, i.e., units of pi.**
> **% actRp and actAs are the realized values of Rp and As from**
> **% the filter design.**
> **% Test calls:**
> **% LVxDesignEquirippHPF(0.2,60,0.45,0.55)**
> **% LVxDesignEquirippHPF(0.02,60,0.4,0.55)**
> **% LVxDesignEquirippHPF(0.5,70,0.15,0.25)**

13. In this project, we'll modify and use the script

$$[T, F, B] = LV\_DetectContTone(A, Freq, RorSS, SzWin, OvrLap, AudSig)$$

to not only identify an interfering steady-state or rising-amplitude sinusoid, but to remove it from the test signal using an equiripple lowpass, highpass, or notch filter designed automatically in response to the list of candidate interfering frequencies generated through analysis of the spectrogram matrix.

Once the filter is designed, filtering is performed using two methods. The first method is to filter the entire test signal in the time domain using the filter coefficients $[b, a]$ and the function $filter$. The second method is to implement convolution in the frequency domain on each frame of the signal, the output signal then being constructed from the frames. To do this, the test signal is first divided into frames with a certain overlap, such as 50%, to form the matrix $TDMat$, and then each frame (i.e., column) of $TDMat$ is filled out with zeros to double its length. The DFT of all frames (columns) is then obtained, forming the matrix $lgFty$, and the analysis proceeds using the nonnegative bins of $lgFty$. Once the interfering frequency or frequencies have been identified and suitable filter coefficients have been determined, an equivalent impulse response, truncated to the same length as the augmented columns of $lgFty$, is generated and its DFT is obtained. Each frame of $lgFty$ is then multiplied by this DFT, and then the real part of the inverse DFT of $lgFty$ is obtained to form a matrix $lgTD$. A matrix $newTD$, consisting of the upper half of the columns of $lgTD$, is then formed. To perform the proper Overlap and Add routine for the frequency domain convolution, the lower half of $lgTD$ is shifted one column to the right and added to the upper half of $newTD$. The frames of $newTD$ are then concatenated with the proper overlap to generate the filtered output signal. DFTs of the test signal and both versions of the output signal (i.e., filtered in the time domain and via the frequency domain) are computed and plotted for comparison. The test signal and both filtered versions are also played through the computer's audio system for audible comparison.

The actual filter design involves using a lowpas filter when the interfering frequency or frequencies lie at or above 0.95π radians, a highpass filter when the interfering frequency or frequencies lie at or below 0.035π radians, or a notch filter otherwise. The necessary design programs can be supplied using the supplied scripts *LVDesignEquirippLPF* and *LVDesignEquirippBPF*, and two scripts written for exercises previous to this one, *LVxDesignEquirippNotch*, and *LVxDesignEquirippHPF*.

```
function [TF,Fnyq,BnSp] = LVx_DetnFiltFIRContTone(A,Freq,...
RorSS,SzWin,OvrLap,AudSig,Rp,As)
% Creates a test signal comprising an audio signal and an
% interfering sinusoid, and attempts to identify the interfering
% tone. The output arguments and the first six input arguments
% are identical to those of the script LVx_DetectContTone
% After analyzing the signal, the script then evaluates the list of
% candidate interfering tones TF to determine if filtering should
% be performed. Frequencies below 80 Hz are eliminated from the
% list as the two audio files contain prominent 60 Hz components,
% and the program is only designed to filter out one main
% spectral component which lies above 80 Hz. Also, the ear's
% sensitivity at low frequencies is very low, and the audibility of
% such tones is often well-masked by other signal components.
% The candidate frequencies above 80 Hz comprise either a single
% frequency or a number of contiguous frequencies(i.e.,lying in
% adjacent bins of the DFT). When a number of contiguous
% candidate frequencies exist, the lower and upper frequency
% bounds are established. If only a single frequency candidate
% exists, upper and lower bounds surrounding it are created so
% that a notch filter of reasonable width can be designed. When the
% interfering frequency or frequencies lie at or below 0.035pi
% radians (about 140 Hz for this exercise), a highpass filter is
% used; when the interfering frequencies lie above 0.95pi
% radians, a lowpass filter is used. Otherwise, a notch filter is used.
% The filter (an equiripple FIR) is designed to have maximum
% passband ripple of Rp dB and minimum stopband attenuation
% of As dB.
% Filtering is performed two ways, first, by time domain
% convolution of the entire test signal with the designed filter,
% and secondly, by frequency domain convolution
% that uses the original analysis matrix, which is doubled in
```

% column length (by addition of zeros to each column) to
% accommodate both the analysis and frequency domain
% convolution. After the columns of this special matrix have
% been multiplied by the DFT of the designed filter's impulse
% response, the IDFT is obtained to form filtered time
% domain signal frames, and the output signal vector is
% reconstructed by concatenating frames with the proper
% overlap, if any.
% The test signal, the time domain filtered version, and the
% frequency domain filtered version are all played out
% through the computer's audio system for comparison.
% Several figures are created in addition to those associated
% with frequency detection per se, including the filter impulse
% response, the filter frequency response, and the test signal
% frequency content after being filtered as one signal vector,
% and after being filtered using the DFT Convolution
% method, reconstructing the signal vector from frames
%
% [T,F,B] = LVx_DetnFiltFIRContTone(0.015,200,0,512,1,1,1,30)
% [T,F,B] = LVx_DetnFiltFIRContTone(0.01,210,0,512,1,1,1,30)
% [T,F,B] = LVx_DetnFiltFIRContTone(0.01,200,0,512,1,2,1,30)
% [T,F,B] = LVx_DetnFiltFIRContTone(0.1,200,0,512,1,3,1,30)
% [T,F,B] = LVx_DetnFiltFIRContTone(0.01,500,0,512,1,1,1,30)
% [T,F,B] = LVx_DetnFiltFIRContTone(0.01,500,0,512,1,2,1,30)
% [T,F,B] = LVx_DetnFiltFIRContTone(0.08,500,0,512,1,3,1,30)
%
% [T,F,B] = LVx_DetnFiltFIRContTone(0.025,200,1,512,1,1,1,30)
% [T,F,B] = LVx_DetnFiltFIRContTone(0.02,200,1,512,1,2,1,30)
% [T,F,B] = LVx_DetnFiltFIRContTone(0.1,200,1,512,1,3,1,30)
% [T,F,B] = LVx_DetnFiltFIRContTone(0.02,500,1,512,1,1,1,30)
% [T,F,B] = LVx_DetnFiltFIRContTone(0.018,500,1,512,1,2,1,30)
% [T,F,B] = LVx_DetnFiltFIRContTone(0.1,500,1,512,1,3,1,30)
% [T,F,B] = LVx_DetnFiltFIRContTone(0.07,150,0,512,1,3,1,30)
%
% [T,F,B] = LVx_DetnFiltFIRContTone(0.005,1000,0,512,0,1,1,30)
% [T,F,B] = LVx_DetnFiltFIRContTone(0.005,1000,0,512,1,1,1,30)

The call

[T,F,B] = LVx_DetnFiltFIRContTone(0.02,100,0,512,0,1,1,40)

results in Figures 2.51–2.53, in addition to those figures created as part of the frequency detection process per se.

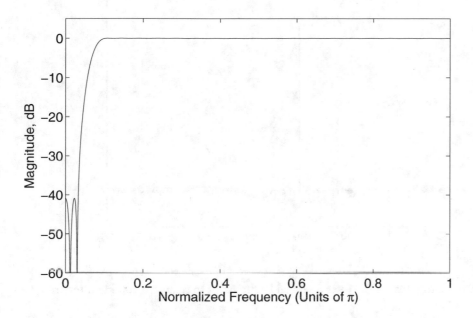

Figure 2.51: Frequency response of the filter designed by the script to remove a 100 Hz tone from the test signal.

The call

[T,F,B] = LVx_DetnFiltFIRContTone(0.01,500,0,512,1,1,1,50)

results in Figures 2.54 and 2.55, in addition to those figures created as part of the frequency detection process per se.

14. Write a script, based on the work done for the script in the previous exercise, that meets the following format and description:

function [T] = LVxAnalyzeModWavFile(strWavFile,SzWin,...
OvrLap,FrqRg,A,Freq,Rp,As,Act,strOutWavFile)
% strWavFile is a .wav file in the target folder (or a folder on
% the MathScript search path to be read and processed
% according the value of Act).
% Act = 1 means analyze only and
% report candidate interfering frequencies as output variable T
% Act = 2 means analyze and filter candidate frequencies if they

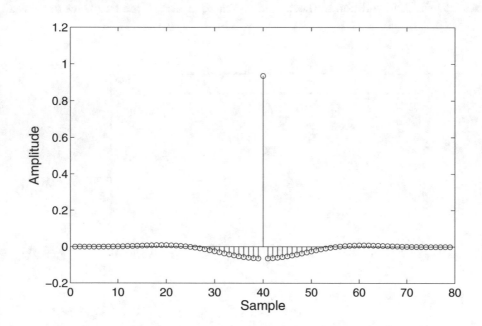

Figure 2.52: Impulse response of the filter designed by the script to remove a 100 Hz interfering tone.

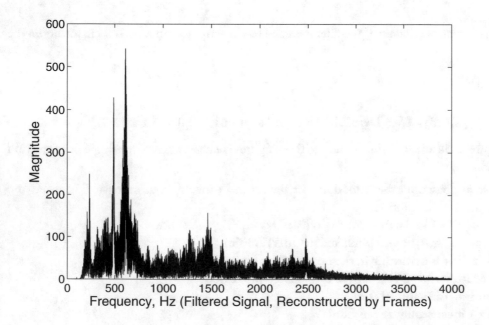

Figure 2.53: Spectrum of filtered output signal reconstructed from frames that were filtered using DFT convolution. Note the attenuated band of frequencies below about 100 Hz.

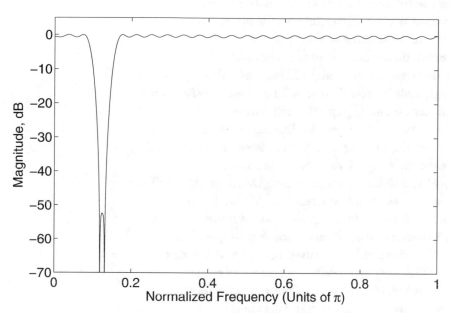

Figure 2.54: Spectrum of 500 Hz notch filtered designed automatically by the script to remove the detected 500 Hz tone.

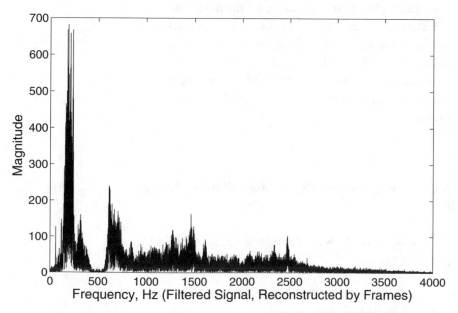

Figure 2.55: Spectrum of filtered output signal reconstructed from frames that were filtered using DFT convolution. Note the notch surrounding 500 Hz.

% lie within the frequency range designated by FrqRg (see below)
% Act = 3 means analyze, filter, and save the filtered output as
% a .wav file under the name strOutWavFile
% Act = 4 means do not analyze or filter, but make a .wav file
% from the test signal, which might have an added component if
% A has magnitude greater than zero. The action may be used to
% generate .wav files having specific frequencies of interfering
% sinusoid.strOutWavFile must be different from strWavFile
% TonFreq, SzWin, OvrLap, A, and Freq have the same meaning
% and use as in the scripts LVx_DetectContTone,
% LVx_DetnFiltFIRContTone, and LVx_DetnFiltIIRContTone
% FrqRg is a vector of two frequencies in Hz that define a range
% over which to consider filtering. This is useful when there are
% multiple interfering frequencies since only simple filters
% (LPF, HPF, and Notch) can be used per call to this script.
% Successive calls specifying different frequency ranges
% will allow removal, for example, of multiple interfering
% frequencies lying at a distance from one another.
%
% Rp and As are the desired maximum passband ripple in dB
% and minimum stopband attenuation in dB for the filter.
% Generally, the minimum value of As should be experimentally
% determined and used; due to masking caused by transient
% signal tones, the audibility of a low level persistent tone can
% often be eliminated using As = 20 dB or 30 dB. Very strong
% persistent tones may require correspondingly higher
% values of As.
%
% Test calls:
%
% [T] = LVx_AnalyzeModWavFile('drwatsonSR8K.wav',512,0,...
[],0,0,1,20,1,[])
%
% [T] = LVx_AnalyzeModWavFile('drwatsonSR8K.wav',512,0,...
[40,80],0,0,1,20,2,[])
%
% [T] = LVx_AnalyzeModWavFile('drwatsonSR8K.wav',512,0,...
[40,80],0,0,1,20,3,'drw8Kless60Hz.wav')
%

% [T] = LVx_AnalyzeModWavFile('drw8Kless60Hz.wav',512,0,...
[40,80],0,0,1,20,1,[])
%
% [T] = LVx_AnalyzeModWavFile('drwatsonSR8K.wav',512,0,...
[40,80],0.02,400,1,20,1,[])
%
% [T] = LVx_AnalyzeModWavFile('drwatsonSR8K.wav',512,0,...
[40,80],0,0,1,20,2,[])
%
% [T] = LVx_AnalyzeModWavFile('drwatsonSR8K.wav',512,0,...
[],0.025,400,1,20,4,'drw8Kplus400Hz.wav')
%
% [T] = LVx_AnalyzeModWavFile('drw8Kplus400Hz.wav',512,0,...
[200,600],0.025,400,1,20,2,[])

After making the call

$$[T] \quad = \quad LVx\_AnalyzeModWavFile('drwatsonSR8K.wav', 512, 0,$$
$$[], 0.025, 400, 1, 20, 4,'drw8Kplus400Hz.wav')$$

which adds a 400 Hz tone to the audio file *'drwatsonSR8K.wav'* and stores the result as *'drw8Kplus400Hz.wav'*, the call

$$[T] \quad = \quad LVx\_AnalyzeModWavFile('drw8Kplus400Hz.wav',...$$
$$512, 0, [], 0.025, 400, 1, 20, 2, [])$$

is made, which displays the minimum energy frame spectrogram and the several post-filtering spectral plots as well as the filter's frequency response (see example Figures 2.56–2.58).

15. In this project, we explore the computational requirements to meet certain FIR design requirements. The goal of the project is to show that under some circumstances, the shortest filter is not the one that is most computationally efficient. Under certain conditions, the Frequency Sampling Realization Method can be employed to greatly reduce computational requirements. To explore this, we'll design a lowpass filter of low passband width ($\omega_p < 0.1\pi$) using the Frequency Sampling Design technique, determine the realized values of R_P, and A_S, implement the filter using the Frequency Sampling Realization Method, and then determine the total computation necessary to generate each sample of output. We'll then design an equivalent equiripple lowpass filter using as the design criteria the values of ω_p, ω_s and the realized values of R_P and A_S determined above. Then we will repeat the exercise for a lowpass filter having $\omega_p = 0.4\pi$.

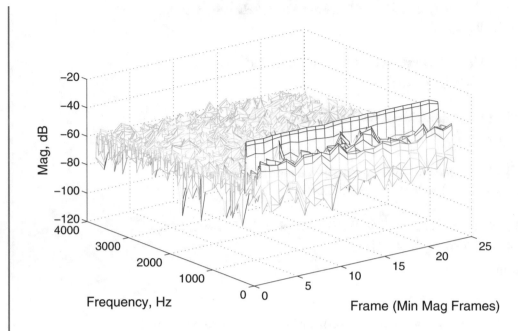

Figure 2.56: Minimum-magnitude-frame spectrogram of signal '*drwatson8Kplus400Hz.wav*'.

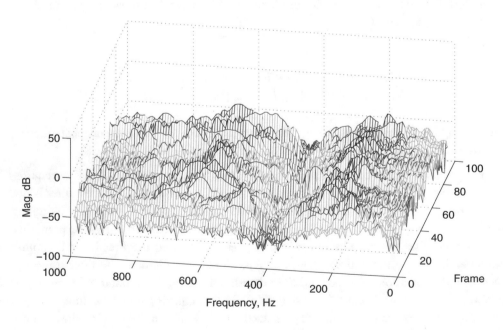

Figure 2.57: Spectrogram (to 1000 Hz, for all frames) of signal '*drwatson8Kplus400Hz.wav*' after being automatically filtered to remove the persistent 400 Hz tone.

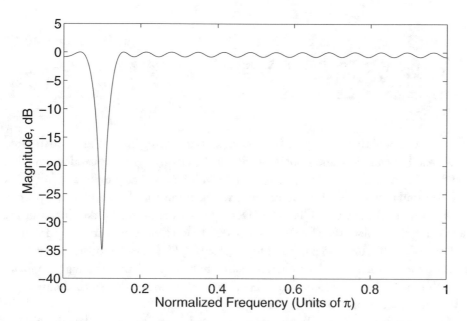

Figure 2.58: Frequency response of equiripple notch filter automatically designed to remove the persistent 400 Hz tone in the sound file 'drwatson8Kplus400Hz.wav'.

(a) Design an FIR using the Frequency Sampling Method, having its zero and positive frequency sample amplitudes valued at

$$[ones(1,3),0.5937402,0.1055273,zeros(1,26)]$$

for a total filter length of 61. Determine the normalized frequencies of ω_p and ω_s, which are assumed to lie on frequency samples just below and above the transition band, respectively. Evaluate the frequency response of the filter and obtain realized values of R_P, and A_S.

(b) Compute the Frequency Sampling realization of the filter designed in (a) and count the number of multiplications and additions needed in the realization to compute each sample of output.

(c) Using ω_p, ω_s, R_P, and A_S, design an equiripple lowpass filter and note its length.

(d) Count the number of additions and multiplications needed to implement the filter designed in (b) (recall that it is a linear phase filter and should be implemented as such to minimize computation).

(e) Compare the results and determine which realized filter is more computationally economic to use.

(f) Repeat steps (a) through (d), starting with an FIR designed using the Frequency Sampling Method having its zero and positive frequency sample amplitudes valued at

$$[ones(1,13),0.592998,0.1095556,zeros(1,15)]$$

C H A P T E R   3

Classical IIR Design

3.1   OVERVIEW

In the previous chapter, we found that it was possible to design filters having linear phase with desired passband ripple, stopband attenuation, and transition widths. Achieving high stopband attenuation along with narrow transition bands (i.e., "steep roll-off"), however, comes at the price of long FIR filters having high computational cost. In this chapter, we examine the design of four types of infinite impulse response filters known as **Classical IIRs,** which were originally developed in the continuous domain using the Laplace transform. Classical IIR design theory permits ready design of recursive digital filters for all the common passband types (lowpass, highpass, bandpass, and notch). Such filters generally do not have linear phase characteristics, but steep roll-offs and high stopband attenuation can be achieved with far less computational overhead than with FIRs. Accordingly, these filters constitute yet another valuable asset in digital signal processing.

In this chapter, we briefly describe the Laplace transform and domain, followed by a description of prototype analog lowpass filters of the four standard Classical filter types (Butterworth, Chebyshev Types I and II, and Elliptic). We then see how the basic lowpass prototype filters can be converted into other filter types, such as highpass, etc. Having thus established the means to design any of the standard passband filter types (lowpass, highpass, bandpass, bandstop) in the analog domain, we then investigate two methods, Impulse Invariance and the Bilinear Transform, to convert an analog filter into an equivalent digital filter. We then describe MathScript's functions for designing Classical IIRs, and conclude the chapter with a brief mention of IIR optimization algorithms with a pointer to further reading.

3.2   LAPLACE TRANSFORM

3.2.1   DEFINITION
Just as the DTFT

$$DTFT(x[n]) = X(e^{j\omega}) = \sum_{n=-\infty}^{\infty} x[n]e^{-j\omega n}$$

is a discrete case or version of the continuous domain Fourier transform

$$F(\omega) = \int_{-\infty}^{\infty} x(t)e^{-j\omega t} dt$$

the z-transform, which we have already studied,

$$X(z) = \sum_{n=-\infty}^{\infty} x[n]z^{-n} \tag{3.1}$$

is actually a discrete form of the Laplace transform, which is defined as

$$\pounds(s) = \int_{-\infty}^{\infty} x(t)e^{-st}dt \tag{3.2}$$

with

$$s = \sigma + j\omega$$

where σ, a real number, is a damping factor, and $\omega = 2\pi f$.

Recall that most sequences we deal with in the real world can be characterized as right-handed, or identically zero for values of time less than zero, which simplifies the z-transform to

$$X(z) = \sum_{n=0}^{\infty} x[n]z^{-n} \tag{3.3}$$

and analogously for right-handed, continuous time signals, the Laplace transform becomes

$$\pounds(s) = \int_{0}^{\infty} x(t)e^{-st}dt \tag{3.4}$$

3.2.2   CONVERGENCE

When the integral in Eq. (3.2) converges (which it does for certain cases) the Laplace Transform is said to exist or to be defined; when the integral does not converge the transform is said to be undefined.

Many signals of interest take exponential form, such as real exponentials, complex exponentials, sine, cosine, etc., and the Laplace transforms for such signals are generally defined.

Example 3.1.   Determine the Laplace transform and convergence criteria for the signal

$$f(t) = e^{at}u(t)$$

with a being, in general, a complex number, and $u(t)$ has the value 0 for $t < 0$ and 1 for all other t. We can determine the Laplace transform as

$$\pounds(f(t)) = \int_{0}^{\infty} f(t)e^{-st}dt = \int_{0}^{\infty} e^{at}e^{-st}dt \tag{3.5}$$

$$= \int_{0}^{\infty} e^{-(s-a)t}dt = \frac{-1}{s-a}\left(e^{-(s-a)t}\Big|_{0}^{\infty}\right)$$

The net result is

$$\pounds\,(e^{at}u(t)) = \frac{1}{s-a} \tag{3.6}$$

where the result can also be obtained from a standard table of Laplace Transforms which can be found in many books, such as Reference [2]. Note that for the integral at (3.5) to converge, the real part of s (i.e, namely σ), must be greater than $\mathrm{Re}(a)$. For example, if $a = -6 + 2j$, then $\sigma > -6$ for (3.5) to converge. Stated differently, but more generally, the real part of s must lie to the right of the real part of the rightmost pole (as graphed in the complex plane) of the system or function $f(t)$ for the Laplace Transform to converge and thus be defined.

Example 3.2.   Determine the Laplace transform and convergence criteria for

$$f(t) = \cos(\omega_0 t)$$

An easy approach is to use Euler's identity and construct the Laplace integral as

$$\pounds\,(f(t)) = \int_0^\infty f(t)e^{-st}\,dt = \int_0^\infty \frac{1}{2}(e^{j\omega_0 t} + e^{-j\omega_0 t})e^{-st}\,dt$$

which results in

$$\frac{1}{2}\left(\frac{1}{s - j\omega_0} + \frac{1}{s + j\omega_0}\right) \tag{3.7}$$

which results in

$$\frac{1}{2}\left(\frac{(s + j\omega_0) + (s - j\omega_0)}{(s - j\omega_0)(s + j\omega_0)}\right) = \frac{s}{s^2 + \omega_0^2}$$

for $\sigma > 0$.

3.2.3   RELATION TO FOURIER TRANSFORM

It can be seen that the waveform that is correlated with the signal $x(t)$ is a complex exponential which may have an amplitude which grows, shrinks, or remains the same over time, depending on the value of σ. When $\sigma = 0$, $e^{-\sigma t} = 1$, and the resultant integral is identical to the Fourier Transform.

The Laplace Transform is thus a transform that includes all the information of the Fourier Transform, plus a good deal more. Not only are correlations done with constant amplitude orthogonal pairs, correlations are also done with decaying and increasing amplitude orthogonal pairs. This kind of transform uncovers more information about a system than can be obtained with the Fourier Transform. Like the z-transform, the Laplace Transform allows the poles and zeros of a system to be identified.

3.2.4   RELATION TO z-TRANSFORM

Note that $x[n]$ (a sampled sequence), used as an input to the z-transform, as shown in Eq. (3.1), is just a discrete or sampled version of the continuous function $x(t)$ used as an input to the Laplace Transform (as shown in Eq. (3.2)). The complex correlator in the z-transform,

$$z^{-n} = (Me^{j2\pi k/N})^{-n} = M^{-n}(e^{-j2\pi k/N})^n = (e^{-j2\pi k/N}/M)^n$$

where M (magnitude) is a real number, k is frequency (a continuous-valued real number), N is sequence length, and n is the sample index, is a discrete or sampled form of the complex exponential $e^{-\sigma t}e^{-j\omega t}$. This is true since by choosing k and M properly, the sequence z^{-n} will form a sampled version of the continuous Laplace correlator e^{-st}.

3.2.5   TIME DOMAIN RESPONSE GENERATED BY POLES

The time domain response generated by a single pole is

$$y(t) = e^{(\sigma_p + j\omega_p)t}$$

where $(\sigma_p + j\omega_p)$ is the value of the pole and t is time, with σ_p a real number and ω_p the radian frequency, $2\pi f_p$.

Figure 3.1 shows three pairs of complex conjugate poles, a first pair (shown in plot (a)), located in the left half-plane, a second pair on the j-Ω axis (plot (c)), and a third pair in the right half-plane (plot (e)). The corresponding time domain responses, all-real, are shown in plots (b), (d), and (f), respectively.

3.2.6   GENERAL OBSERVATIONS

Values of the Laplace Transform are, in general, complex, and hence they are graphed in the complex plane. The real part of the Laplace Transform represents the damping factor σ, values of which are found along the real (horizontal) axis, and the imaginary part represents frequency, values of which are found along the imaginary (vertical) axis. Both axes range from negative to positive infinity.

- A pole in the left half-plane (i.e., having its real part less than zero) of the s-domain is stable, and corresponds to (or generates) a decaying exponential.

- A pole lying on the j-Ω axis (i.e., having its real part equal to zero) generates a constant, unity-amplitude exponential in response to an impulse and is borderline unstable.

- A pole lying in the right half-plane (i.e., having its real part greater than zero) of the s-Domain represents an exponential that grows with time and hence corresponds to an unstable transfer function.

- The Laplace Transform converges (and is thus defined) for values of s lying to the right of the right-most system pole as graphed in the s-domain. This is analogous to the situation in

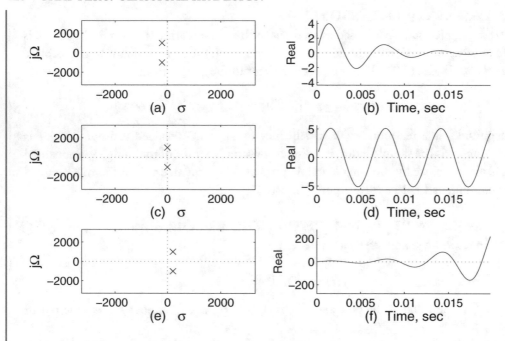

Figure 3.1: (a) A pair of complex conjugate poles in the left half-plane; (b) The time domain response arising from the poles shown in (a); (c) A pair of complex conjugate poles on the imaginary axis; (d) The time domain response arising from the poles shown in (c); (e) A pair of complex conjugate poles in the right half-plane; (f) The time domain response arising from the poles shown in (e).

the z-domain, in which the z-transform (for right-handed sequences) converges for values of z that have magnitude greater than the system pole having the largest magnitude.

The Laplace transform is used in a similar manner to the z-transform, that is, to represent systems as a product of factors containing zeros (in the numerator) divided by a product of factors containing poles (in the denominator), or alternately as a ratio of polynomial expressions in the variable s (rather than z as used in the z-transform). These expressions (i.e., s-domain transfer functions) have similar properties to those of the z-transform, in that, for example, convolution of two time domain signals can be performed equivalently by multiplying the Laplace transforms of the two signals. Many other properties of the Laplace and z- transforms are also analogous.

Laplace transforms generally exist for derivatives of time functions, and thus the Laplace transform is often used to solve problems involving differential equations. The equivalent thing in the discrete domain, of course, is the difference equation.

An in-depth discussion of the Laplace transform is beyond the scope of this book; there are many excellent books available which have covered this subject thoroughly. Our main interest in the Laplace transform is that the standard or classical IIR types (the Butterworth, Chebyshev, and Elliptical filters) originated in the continuous domain and the design theory and procedures are well-developed. By first designing such filters using the Laplace transform, and then using a technique for converting or mapping s-domain values to z-domain values, digital IIR filters can be efficiently designed.

3.3   PROTOTYPE ANALOG FILTERS

3.3.1   NOTATION

The specifications for analog filters differ somewhat from those of the FIRs we have studied so far; the specifications for analog filters are generally given relative to the magnitude squared of the transfer function. Passband ripple is specified by a ripple parameter ϵ, and stopband attenuation by A. Cutoff frequencies are notated using Ω_p and Ω_s (both in radians per second). For a lowpass filter, for example, the specifications would be given as

$$1/(1 + \epsilon^2) \leq |H(j\Omega)|^2 \leq 1 \quad |\Omega| \leq \Omega_p$$
$$0 \leq |H(j\Omega)|^2 \leq 1/A^2 \quad \Omega_s \leq |\Omega|$$

Figure 3.2 illustrates a typical analog filter specification.

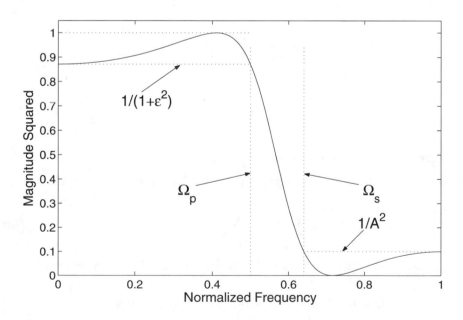

Figure 3.2: The specifications for an analog lowpass filter.

The parameters A and ϵ are related to A_s and R_p according to the following formulas:

$$R_p = -10 \log_{10}(1/(1 + \epsilon^2))$$

$$\epsilon = \sqrt{10^{R_p/10} - 1}$$

$$A_s = -10 \log_{10}(1/A^2)$$

$$A = 10^{A_s/20}$$

In regard to the magnitudes of passband ripple δ_1 and stopband attenuation δ_2, we note that

$$\frac{1 - \delta_1}{1 + \delta_1} = \sqrt{\frac{1}{1 + \epsilon^2}}$$

and therefore

$$\epsilon = \frac{2\sqrt{\delta_1}}{1 - \delta_1}$$

Similarly,

$$\frac{\delta_2}{1 + \delta_1} = \frac{1}{A}$$

and therefore

$$A = \frac{1 + \delta_1}{\delta_2}$$

3.3.2   SYSTEM FUNCTION AND PROPERTIES

An s-domain system function in Direct form is expressed as the ratio of two polynomials in s:

$$H(s) = \sum_{m=0}^{M} b_m s^m / \sum_{n=0}^{N} a_n s^n \tag{3.8}$$

To obtain the system function of a filter specified by its magnitude-squared we note that

$$|H(j\Omega)|^2 = H(j\Omega)H^*(j\Omega) = H(j\Omega)H(-j\Omega) = H(s)H(-s)|_{\Omega=s/j}$$

The above leads to the following observations:

- Poles exhibit mirror-symmetry about the $j\Omega$ axis, never lie on the imaginary axis, and lie on the real axis only for odd order filters.

- For real filters, poles and zeros must either lie on the real axis or exist in complex conjugate pairs.

- A stable and causal filter can be constructed by using only poles in the left half-plane. Zeros are chosen to be minimum-phase, i.e., lying on or to the left of the $j\Omega$ axis. For the filters we will study, the zeros, if any, all lie on the $j\Omega$ axis.

- $H(s)$ can then be constructed from the chosen left half-plane poles and zeros.

Since the Classical IIRs we will study produce pairs of complex conjugate poles, plus a single real pole when the order is odd, a convenient way to express the system function is as the product of biquad sections (and one first order section when the order is odd), each of which is the ratio of two quadratic polynomials is s, having real-only coefficients. The resulting system function is of the form, for even order N

$$H(s) = \Pi_{i=1}^{N/2} \frac{B_{2,i}s^2 + B_{1,i}s + B_{0,i}}{A_{2,i}s^2 + A_{1,i}s + A_{0,i}}$$

where the coefficients B and A are real coefficients for the i-th biquad section. For odd order N, the system function is of the form

$$H(s) = (\Pi_{i=1}^{(N-1)/2} \frac{B_{2,i}s^2 + B_{1,i}s + B_{0,i}}{A_{2,i}s^2 + A_{1,i}s + A_{0,i}})(\frac{1}{s + A_{0,(N+1)/2}})$$

Example 3.3. A certain filter has three poles, namely $[-0.2273 \pm j0.9766, -0.5237]$ and two zeros: $[\pm j2.7584]$; write the system function using biquad sections as well as in Direct Form.

To get Direct Form, we multiply the various zero and pole factors (this can be done using the function *poly*, for example)

$$H(s) = \frac{(s - j2.7584)(s + j2.7584)}{(s + 0.2273 - j0.9766)(s + 0.2273 + j0.9766)(s + 0.5237)}$$

which yields

$$H(s) = \frac{s^2 + 7.6088}{s^3 + 0.9783s^2 + 1.2435s + 0.5265}$$

and thus we have $b = [1,0,7.6088]$ and $a = [1,0.9783,1.2435,0.5265]$.

To get the cascaded-biquad form, we combine the two complex conjugate poles into one factor, as well as the two complex conjugate zeros and get a single biquad section, with one first order section:

$$H(s) = (0.0692)(\frac{s^2 + 7.6089}{s^2 + 0.4546s + 1.0054})(\frac{1}{s + 0.5237}) \qquad (3.9)$$

To convert an s-domain transfer function in Direct Form to one in Cascade Form, use the script

$$[Bbq, Abq, Gain] = LVDirToCascadeClassIIR(b, a, gain)$$

which is similar to the script

$$[Bbq, Abq, Gain] = LVDirToCascade(b, a)$$

presented in the discussion on filter topology in the chapter on the z-transform (found in Volume II of the series; see the Preface to this volume for information on Volume II), except that, in *LVDirToCascadeClassIIR*, the additional input parameter *gain* is used to scale the output parameter *Gain*.

We can verify the correctness of Eq. (3.9), for example, by running the following m-code:

b = [1,0,7.6088]; a = [1,0.9783,1.2435,0.5265];
[Bbq,Abq,Gain]=LVDirToCascadeClassIIR(b,a,1)

which yields

Bbq = [1, 0, 7.6088]; Abq = [1,0.4546,1.0054; 0,1,0.5237]; Gain = 1

To convert from Cascade to Direct Form, use the script

$$[b, a, k] = LVCas2DirClassIIR(Bbq, Abq, Gain)$$

3.3.3   COMPUTED FREQUENCY RESPONSE

To obtain the frequency response of an analog filter, we can use the function $polyval(p, x)$, which evaluates the value of a polynomial, the coefficients of which are p, at frequency x. By making x a vector of desired frequencies, and evaluating both the b and a polynomials of a filter, and taking the ratio, we obtain the frequency response. In the function

$$H = LVsFreqResp(b, a, HiFreqLim, FigNo)$$

the argument $HiFreqLim$ is the high limit frequency at which to evaluate the response of the filter defined by its numerator and denominator polynomial coefficients b and a, respectively, and $FigNo$ is the number to assign to the figure created for the plots.

```
function H = LVsFreqResp(b,a,HiFreqLim,FigNo)
FR = (0:HiFreqLim/5000:HiFreqLim); s = j*FR;
H = polyval(b,s)./polyval(a,s);
figure(FigNo); subplot(311); yplot = 20*log10(abs(H)+eps);
plot(FR,yplot); xlabel('(a) Freq, Rad/s'); ylabel('Mag, dB')
axis([0 HiFreqLim -100 5])
subplot(312); plot(FR,abs(H)); xlabel('(b) Freq, Rad/s');
```

```
ylabel('Mag'); axis([0 HiFreqLim 0 1.05]);
subplot(313); plot(FR,unwrap(angle(H)))
xlabel('(c) Freq, Rad/s'); ylabel('Radians')
axis([0 HiFreqLim -inf inf])
```

Example 3.4.   Evaluate the frequency response of a fifth-order analog lowpass Butterworth filter having a frequency cutoff of π radians.

We'll study Butterworth filters in detail shortly, but for the moment we can obtain a lowpass analog design using the m-code shown, which is followed with a call to *LVsFreqResp* to plot the net frequency response, which results in Fig. 3.3.

```
[b,a] = butter(5,pi,'s')
H = LVsFreqResp(b,a,2*pi,7)
```

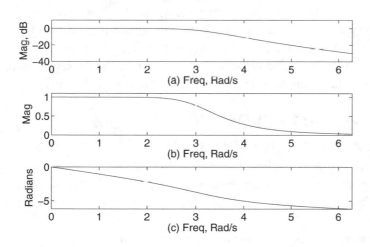

Figure 3.3: (a) Frequency response in dB to frequency 2π radians (1 Hertz) for a fifth order Butterworth lowpass filter; (b) Same, with linear scale rather than dB; (c) Phase response.

Another useful function is one to compute, for an analog lowpass filter, the actual or realized values of R_P and A_S. The input arguments required are the frequency response vector H computed, for example, by the script *LVsFreqResp*, the high frequency limit *HiFreqLim* used to compute H, and the desired or user-specified band limits, Ω_P and Ω_L. As we proceed through the chapter, we will develop scripts that design lowpass filters with user-specified band limits, and hence this script will prove useful for evaluating the actual realized filter performance using the given band limits. For filters other than lowpass, we will, as we proceed through the chapter, write scripts specific to the passband type to evaluate actual filter performance.

```
function [NetRp,NetAs] = LVsRealizedFiltParamLPF(H,OmgP,...
OmgS,HiFreqLim)
Lfr = length(H); mFr = HiFreqLim;
Lenpb = round(OmgP/mFr*Lfr); pb = H(1:Lenpb);
mnpb = min(abs(pb)); NetRp = 20*log10(1/mnpb);
sbStrt = (OmgS/mFr); sb = H(round(sbStrt*Lfr):Lfr);
maxsb = max(abs(sb)); NetAs = 20*log10(1/maxsb);
```

3.3.4   GENERAL PROCEDURE FOR ANALOG/DIGITAL FILTER DESIGN

Our general procedure will be to design prototype lowpass filters based on the well-known analog filters discussed in detail below. Other analog filter types such as highpass, bandpass, and bandstop can be designed by first designing a prototype lowpass analog filter and then using a variable transform or substitution which will convert a lowpass filter transfer function to a different filter type such as highpass, etc. To obtain the desired digital filter, an analog-to-digital transform (such as the Bilinear transform, for example) can then be used to convert the design to the digital domain. An alternate method is to design a lowpass analog filter, convert it into a digital filter, and then use a digital variable-substitution method to convert a lowpass prototype digital filter into another digital passband type, such as highpass, etc. Our discussion below will concentrate on the former method, designing all passband types in the analog domain and then converting to the digital domain.

3.4   ANALOG LOWPASS BUTTERWORTH FILTERS

3.4.1   DESIGN BY ORDER AND CUTOFF FREQUENCY

The Butterworth filter characteristic is one of maximal flatness in the passband at frequency zero, and maximum flatness in the stopband at infinite frequency, which is a desirable trait in certain applications, such as audio amplifiers. The transition band roll-off rate, however, is very shallow for a given filter order N.

The desired magnitude-squared frequency response of a Butterworth filter is

$$|H(j\Omega)|^2 = \frac{1}{1 + (\Omega/\Omega_C)^{2N}}$$

where Ω_C is the cutoff frequency in radians per second, and N is the order of the filter.

The following script can be used (by changing the values of N and *NormFrq* as desired) to plot the frequency response for any desired range of N and normalized frequency (Ω/Ω_C):

```
function LVButterMagSqCurves
figure(79); NormFrq = 0:0.01:2;
for N = 1:1:12; hold on
H = 1./(1 + (NormFrq).^(2*N));
plot(NormFrq,H,'-'); end; hold off
xlabel('Frequency, Units of \Omega_C')
```

ylabel('Magnitude')

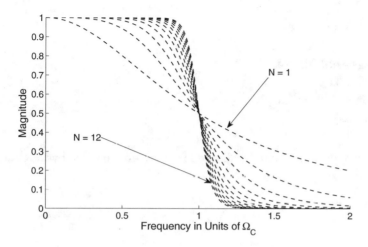

Figure 3.4: The frequency response curves for Butterworth filters of orders 1 to 12.

The system function will be

$$H(s)H(-s) = |H(j\Omega)|^2 = \frac{1}{1 + (s/j\Omega_C)^{2N}} = \frac{(j\Omega_C)^{2N}}{s^{2N} + (j\Omega_C)^{2N}}$$

The roots of the denominator (i.e., the poles of the transfer function) may be found using the expression

$$p_k = (\Omega_C)e^{j(\pi/2N)(2k+N+1)} \tag{3.10}$$

where $k = 0{:}1{:}2N\text{-}1$ and the system function can be written as

$$H(s) = \frac{(\Omega_C)^N}{\Pi(s - p_k)} \tag{3.11}$$

where only the p_k lying in the left half-plane are used. The numerator of Eq. (3.11) normalizes the filter gain to 1.0 at frequency 0.0 rad/sec.

Example 3.5.   Determine the system function for a Butterworth filter having $\Omega_C = 2$ and $N = 3$.

The following code produces all the possible poles for the system function; however, to ensure stability, only those poles lying in the left half-plane of the s-domain are used. The code plots the poles chosen from the left half-plane and the circle upon which they lie. The system function can then be generated using Eq. (3.11) and the poles obtained as *NetP* in the code below.

```
function LVButterPoles(N,OmegaC)
% LVButterPoles(3,2)
k = 0:1:2*N-1;
P = OmegaC*exp(j*(pi/(2*N))*(2*k+N+1))
NetP = P(find(real(P)<0))
figure(90); clf; hold on; args = 0:0.02:2*pi;
plot(real(NetP),imag(NetP),'bx');
xlabel('Real'); cnums = OmegaC*exp(j*args);
plot(real(cnums),imag(cnums),':')
ylabel('Imaginary')
```

Figure 3.5 is a plot of the three poles computed by the code above, accompanied by the circle upon which they lie in the s-plane.

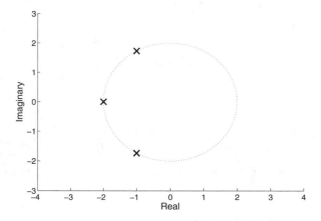

Figure 3.5: The three poles for a third order Butterworth filter. To ensure stability, only the left half-plane poles of the possible six poles are used (right half-plane poles are not shown).

- The poles of a Butterworth filter all lie on a circle of radius Ω_C.

- None of the poles lie on the imaginary axis.

- All complex poles are accompanied by their complex conjugates.

- To ensure stability, only poles to the left of the imaginary axis in the s-plane are used.

Example 3.6. Evaluate the frequency response of the system function of a Butterworth lowpass filter having three poles and a cutoff frequency of 1.0 rad/s by direct evaluation of the transfer function, factor by factor. Check the results using the script *LVsFreqResp*.

The following code computes $Net\ P$ and then obtains the product of the factors of the transfer function over a range of normalized test frequencies, each factor magnitude being of the form

$$\frac{1}{s - p_k}$$

The frequency response is normalized by Ω^N so that the response at normalized $\Omega = 0$ is 1.0.

```
function LVButterFR(N,OmegaC)
% LVButterFR(3,1)
k = 0:1:2*N-1;
P = OmegaC*exp(j*(pi/(2*N))*(2*k+N+1));
NetP = P(find(real(P)<0)); Freq = (0:0.01:4*OmegaC);
s = j*Freq; Resp = OmegaC^N; for Ctr = 1:1:length(NetP)
Resp = Resp.*(1./(s - NetP(Ctr))); end; figure(97);
subplot(311); plot(Freq,20*log10(abs(Resp)));
xlabel('Freq, Rad/s'); ylabel('Mag, dB')
subplot(312); plot(Freq,abs(Resp)); xlabel('Freq, Rad/s');
ylabel('Mag'); subplot(313); plot(Freq,unwrap(angle(Resp)))
xlabel('Freq, Rad/s'); ylabel('Radians')
```

The result from running the code above is shown in Fig. 3.6. Note that the phase response of the Butterworth filter is reasonably linear within the passband.

We can check the work above using the following script, *LVButterFRViaPoly*, which uses the function *LVsFreqResp* (introduced earlier) to obtain the frequency response; the results are shown in Fig. 3.7.

```
function LVButterFRViaPoly(N,OmegaC,HiFreqLim)
% LVButterFRViaPoly(3,1,4)
k = 0:1:2*N-1;
P = OmegaC*exp(j*(pi/(2*N))*(2*k+N+1))
NetP = P(find(real(P)<0)); b = 1; a = poly(NetP);
H = LVsFreqResp(b,a,HiFreqLim,12);
```

MathScript provides a function to compute the poles of a normalized (i.e., $\Omega_C = 1.0$ rad/s) Butterworth lowpass filter. The syntax is

$$[z, p, K] = buttap(N)$$

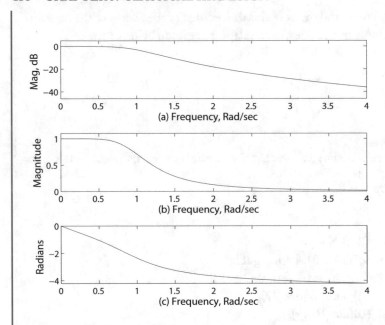

Figure 3.6: (a) Frequency response in dB of a Butterworth filter of order 3, up to a frequency equal to four times the cutoff frequency of 1 rad/s; (b) Same in linear units; (c) Phase response of same.

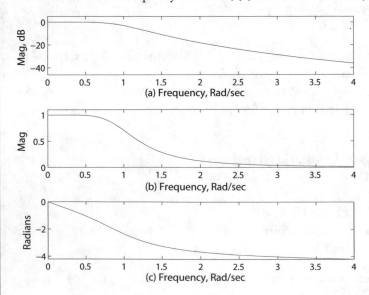

Figure 3.7: (a) The magnitude in dB of the frequency response of a 3-rd order Butterworth filter having $\Omega_C = 1$; (b) Same in linear units; (c) Phase response of same.

where z is the vector of (finite) zeros (returned as empty for Butterworth filters), p is the vector of poles, K is the gain (1.0) necessary to obtain unity magnitude response at $\Omega = 0$ rad/s, and N is the desired order.

Example 3.7. Obtain the poles of a 3-rd order lowpass Butterworth filter, scale them for $\Omega_C = 3$ rad/s, and compute the necessary value of K for the filter having $\Omega_C = 3$ rad/s. Compute and write the system function in Direct and Cascade forms and check the frequency response.

We make the call

$$[z,p,k] = buttap(3)$$

to obtain the normalized poles $p = [-0.5 \pm j0.866, -1.0]$, which are then multiplied by the desired Ω_C (3 rad/s) to yield the poles for $\Omega_C = 3$ rad/s as $P = [-1.5 \pm j2.598, -3.0]$; from k, returned as 1.0, we obtain the new value of K as $k(\Omega_C^N) = (1.0)(3^3) = 27$.

The system function is

$$H(s) = \frac{27}{(s + 1.5 + j2.598)(s + 1.5 - j2.598)(s + 3)}$$

We can check the frequency response with a simple script:

b = 27; a = poly([(-1.5 + j*2.598),(-1.5 - j*2.598),-3]);
H = LVsFreqResp(b,a,20,13);

A script that produces the system function in Direct Form as well as Cascade Form is

function LVButterPolesAndSysFcn(N,OmC)
% LVButterPolesAndSysFcn(3,3)
[z,p,k] = buttap(N),
P = OmC*p, K=k*OmC^N,
b = real(poly(z)), a = real(poly(P)),
[Bbq,Abq,Gain] = LVDirToCascadeClassIIR(b,a,K)

which yields $b = 1$, $a = [1,6,18,27]$, $B\_bq = 1$, $A\_bq = [1,3,9; 0,1,3]$, and $Gain = 27$, and the Direct and Cascade system functions as

$$H(s) = \frac{1}{s^3 + 6s^2 + 18s + 27} = \frac{1}{(s^2 + 3s + 9)(s + 3)}$$

Note that letting $s = 0$ (i.e., evaluating the transfer function at DC or frequency zero) yields at output of 1/27, so to achieve unity gain at DC, include $Gain$ with the system transfer function:

$$H(s) = \frac{27}{s^3 + 6s^2 + 18s + 27} = \frac{27}{(s^2 + 3s + 9)(s + 3)}$$

3.4.2   DESIGN BY STANDARD PARAMETERS

To design a Butterworth lowpass filter to meet the standard specifications of Ω_P, Ω_S, R_P, and A_S, it is necessary to determine the required value of N. We start by noting that at Ω_P, the magnitude of response should be R_P, and at Ω_S, the magnitude of response should be A_S:

$$- 10\log_{10}(1/(1 + (\Omega_P/\Omega_C)^{2N})) = R_P \tag{3.12}$$

and

$$- 10\log_{10}(1/(1 + (\Omega_S/\Omega_C)^{2N})) = A_S \tag{3.13}$$

After solving for N, we get

$$N = \frac{\log_{10}[(10^{R_P/10} - 1)/(10^{A_S/10} - 1)]}{2\log_{10}(\Omega_P/\Omega_S)} \tag{3.14}$$

where N (usually not an integer as computed above) must be rounded up to the next integer to ensure that the specifications are met.

Once N has been obtained, it can be used in either of Eqs. (3.12) or (3.13) to solve for values of Ω_C to satisfy either R_P exactly at Ω_P, or A_S exactly at Ω_S. The resultant formulas, respectively, are

$$\Omega_{C1} = \frac{\Omega_P}{\sqrt[2N]{10^{R_P/10} - 1}} \tag{3.15}$$

$$\Omega_{C2} = \frac{\Omega_S}{\sqrt[2N]{10^{A_S/10} - 1}} \tag{3.16}$$

The range of acceptable values for Ω_C is obtained by solving for Ω_C in both of Eqs. (3.15) and (3.16) and choosing Ω_C between Ω_{C1} and Ω_{C2}.

Example 3.8.   Design a Butterworth Filter having $\Omega_P = 0.4\ Rad/s$, $\Omega_S = 0.7\ Rad/s$, $R_P = 0.1$ dB, and $A_S = 40$ dB.

The following code computes N, Ω_C for both cases, the average of the two values of Ω_C, and the poles for the Butterworth filter, and then plots the frequency and phase responses. The result is shown in Fig. 3.8.

```
function [Z,P,K] = LVDesignButterworth(OmgP,OmgS,Rp,As)
% [Z,P,K] = LVDesignButterworth(0.4,0.7,0.2,40)
num = 10^(Rp/10)-1; denom = 10^(As/10)-1;
N = ceil(log10(num/denom)/(2*log10(OmgP/OmgS)))
OmcP = OmgP/((10^(Rp/10)-1)^(0.5/N))
OmcS = OmgS/((10^(As/10)-1)^(0.5/N))
```

OmgC = (OmcP + OmcS)/2; k = 0:1:2*N-1;
P = OmgC*exp(j*(pi/(2*N))*(2*k+N+1));
P = P(find(real(P)<0)); Z = [];
K = OmgC^N;

The following code will obtain the Butterworth poles and zeros, obtain and plot the frequency response, the realized values of R_P and A_S, and the cascade coefficients.

Rp = 0.2; As = 40; OmgP = 0.4; OmgS = 0.7;
[Z,P,K] = LVDesignButterworth(OmgP,OmgS,Rp,As)
H = LVsFreqResp(K*poly(Z),poly(P),4*OmgS,14)
[NetRp,NetAs] = LVsRealizedFiltParamLPF(H,OmgP,...
OmgS,4*OmgS)
[Bbq,Abq,Gain] = LVDirToCascadeClassIIR(poly(Z),poly(P),K)

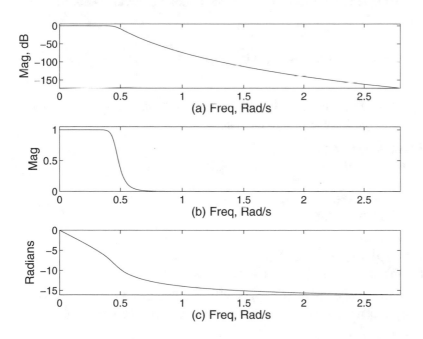

Figure 3.8: (a) Magnitude in dB of frequency response of a Butterworth filter having Ω_P = 0.4 rad/s, Ω_S = 0.7 rad/s, R_P = 0.2 dB, and A_S = 40 dB. The necessary value of N was computed to be 11; (b) Same, but in linear scale; (c) Phase response.

3.5   LOWPASS ANALOG CHEBYSHEV TYPE-I FILTERS

Chebyshev filters are based on the Chebyshev polynomials, which are defined as

$$T_N(x) = \begin{cases} \cos(N\cos^{-1}(x)) & |x| \le 1 \\ \cosh(N\cosh^{-1}(x)) & |x| > 1 \end{cases}$$

The Chebyshev Type-I filter has an equiripple characteristic in the passband, and decreases monotonically in the stopband; the Chebyshev Type-II filter is monotonic in the passband and equiripple in the stopband.

3.5.1   DESIGN BY ORDER, CUTOFF FREQUENCY, AND EPSILON

The magnitude squared function for a **Chebyshev Type I** filter is defined as:

$$|H(\Omega)|^2 = \frac{1}{1 + \epsilon^2(T_N(x))^2}$$

where $x = \Omega/\Omega_P$. The parameter ϵ determines, for a given N, the tradeoff between passband ripple and transition band steepness.

Example 3.9.   Compute and display the magnitude squared function for a Type-I Chebyshev filter having $\epsilon = 0.4$ and $N = 5$.

Note in the following code that the normalized frequency range is broken into two subranges to accommodate the two functions (cosine and hyberbolic cosine).

```
function LVCheby1MagSquared(Ep,N)
% LVCheby1MagSquared(0.4,5)
inc = 0.01; xLo = 0:inc:1;
xHi = 1+inc:inc:3; TnLo = cos(N*acos(xLo));
TnHi = cosh(N*acosh(xHi)); T = [TnLo,TnHi];
MagHSq = 1./(1 + Ep^2*(T.^2));
figure; xplot = [xLo, xHi]; plot(xplot,MagHSq);
xlabel('Norm Freq (\Omega/\Omegac)'); ylabel('Mag Squared')
```

Figure 3.9 shows the magnitude squared of a Chebyshev Type I filter having $N = 4$ for several values of ϵ. As ϵ decreases, the passband ripple decreases but the transition band becomes wider, i.e., the roll-off is less steep.

The ripple characteristics for Type-I Chebyshev filters differ according to whether N is even or odd. This is illustrated in Fig. 3.10.

$$|H(j0)|^2 = 1 \qquad |H(j1)|^2 = 1/(1 + \epsilon^2) \quad N \text{ odd}$$
$$|H(j0)|^2 = 1/(1 + \epsilon^2) \quad |H(j1)|^2 = 1/(1 + \epsilon^2) \quad N \text{ even}$$

The poles of the system function are the roots of

$$1 + \epsilon^2(T_N(s/j\Omega_C))^2$$

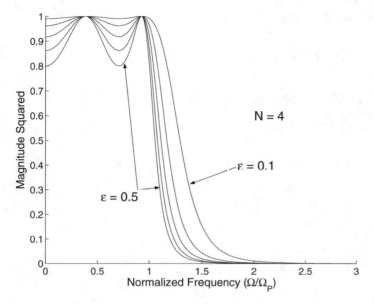

Figure 3.9: The magnitude squared function for a Chebyshev Type I filter having $N = 4$ for values of ϵ equal to 0.1, 0.2, 0.3, 0.4, and 0.5. With $\epsilon = 0.5$, the passband ripple is the largest, but the transition band is the narrowest.

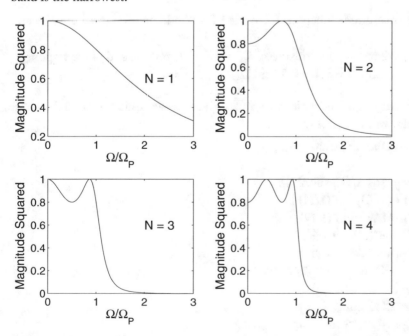

Figure 3.10: The magnitude squared characteristics of a Chebyshev Type-I lowpass filter having $\epsilon = 0.5$ for several even and odd values of N.

and the left half-plane poles $p_k = \sigma_k + j\Omega_k$ can be computed as

$$\sigma_k = (A\Omega_C)\cos[\pi/2 + \pi(2k + 1)/2N]$$

$$\Omega_k = (B\Omega_C)\sin[\pi/2 + \pi(2k + 1)/2N]$$

where k = 0:1:N-1 and

$$A = (\sqrt[N]{\alpha} - \sqrt[N]{1/\alpha})/2$$

$$B = (\sqrt[N]{\alpha} + \sqrt[N]{1/\alpha})/2$$

with

$$\alpha = 1/\epsilon + \sqrt{1 + 1/\epsilon^2}$$

The system function is

$$H(s) = \frac{K}{\Pi(s - p_k)}$$

where K is chosen so that the magnitude function at Ω = 0 is 1 for N odd or $1/\sqrt{1 + \epsilon^2}$ for N even.

Example 3.10.   Write a script that computes the poles, K, and the magnitude and phase responses for a Type-I Chebyshev filter having N = 5, ϵ = 0.5, and Ω_C = 1 rad/sec.

A straightforward application of the various formulas given above results in the following code, the result of which is shown in Fig. 3.11.

```
function LVCheby1(N,OmC,Epsilon)
% LVCheby1(5,1,0.5)
Alpha = 1/Epsilon + sqrt(1 + 1/(Epsilon^2));
B = 0.5*( Alpha^(1/N) + (1/Alpha)^(1/N) );
A = 0.5*( Alpha^(1/N) - (1/Alpha)^(1/N) );
k = 0:1:N-1; arg = pi/2 + pi*(2*k + 1)/(2*N);
SigK = A*cos(arg); OmK = B*sin(arg);
P = OmC*(SigK + j*OmK); K = prod(abs(P));
if rem(N,2)==0 % N even
K = 1/sqrt(1 + Epsilon^2)*K; end;
H = LVsFreqResp(K,poly(P),2*OmC,8);
```

An alternate computational method for the poles of a Type-I Chebyshev filter, which produces the same result as the previously described method, is as follows. Compute

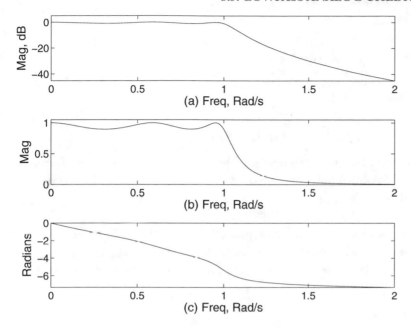

Figure 3.11: (a) Magnitude response in dB of a Chebyshev Type-I filter having N -5, $\epsilon = 0.5$, and Ω_C = 1 rad/sec.; (b) Magnitude response (linear) of same; (c) Phase response of same.

$$v_0 = \sinh^{-1}(1/\epsilon)/N \tag{3.17}$$

and

$$k = -(N-1) : 2 : (N-1)$$

and then

$$P_K = -\sinh(v_0)\cos(k\pi/2N) + \cosh(v_0)\sin(k\pi/2N) \tag{3.18}$$

This method assumes $\Omega_C = 1$, so the poles P_K must be scaled by the actual value of Ω_C.

Example 3.11.   For the example above, in which $\epsilon = 0.5$, $\Omega_C = 1$, and $N = 5$, compute the poles using the alternate method just described.

To compare results with the previous method, run the code from the previous example to display the pole values, and then run the code below.

```
function p = LVCheby1Poles2ndMethod(N,OmC,Epsilon)
% p = LVCheby1Poles2ndMethod(5,2,0.5)
```

k = -(N-1):2:N-1; r = k*pi/(2*N);
v0 = asinh(1/Epsilon)/N;
p = OmC*(-sinh(v0)*cos(r) + j*cosh(v0)*sin(r));

MathScript provides a function to compute the poles of a normalized (i.e., Ω_C = 1.0 rad/s) Chebyshev Type-I lowpass filter. The syntax is

$$[z, p, k] = cheb1ap(N, Rp)$$

where z is the vector of (finite) zeros (returned as empty for Chebyshev Type-I filters), p is the vector of poles, k is the gain necessary to obtain unity magnitude response at Ω = 0 rad/s, N is the desired order, and Rp is the passband ripple in dB.

Example 3.12.   Obtain the poles of a second order lowpass Chebyshev Type-I filter having R_P = 0.2 dB, scale them for Ω_C = 4 rad/s, and compute the necessary value of K for the filter having Ω_C = 4 rad/s. Write the system function. Compute the Cascade Form coefficients, and convert them back to Direct Form.

We make the call

$$[z,p,k] = cheb1ap(2,0.2)$$

to obtain the normalized poles p = [-0.9635 $\pm$ j1.1952], which are then multiplied by the desired Ω_C (4 rad/s) to yield the poles for Ω_C = 4 rad/s as P = [-3.8542 $\pm$ j4.7807]; from the normalized value of k (returned from the call above as 2.3032), the new value of K is obtained as $k(\Omega_C^N)$ = (2.3032)(4²) = 36.85.

A script to perform these computations and produce the system function is

function LVCheby1PolesAndSysFcn(N,OmC,Rp)
% LVCheby1PolesAndSysFcn(2,4,0.2)
[z,p,k] = cheb1ap(N,Rp),
P = OmC*p, K = k*OmC^N,
b = real(poly(z)), a = real(poly(P)),
H = LVsFreqResp(K,poly(P),2*OmC,19);
[Bbq,Abq,Gain] = LVDirToCascadeClassIIR(b,a,K)
[b,a,k]=LVCas2DirClassIIR(Bbq,Abq,Gain)

The call

LVCheby1PolesAndSysFcn(2,4,0.2)

yields b = 1 and a = [1,7.7083,37.709], which are the same as Bbq and Abq since there are only two poles.

The system function is

$$H(s) = \frac{36.85}{s^2 + 7.7083s + 37.7093}$$

3.5.2   DESIGN BY STANDARD PARAMETERS

The parameters N, ϵ, and Ω_C are used to design a Chebyshev Type-I filter. From the input parameters Ω_P and Ω_S we can immediately compute ϵ and A (stopband attenuation in dB) as

$$\epsilon = \sqrt{10^{R_P/10} - 1}$$

and

$$A = 10^{A_S/20}$$

We note that

$$\Omega_C = \Omega_P$$

and we define

$$\Omega_T = \Omega_S / \Omega_P$$

Then

$$N = (\frac{\log_{10}(g + \sqrt{g^2 - 1})}{\log_{10}(\Omega_T + \sqrt{\Omega_T^2 - 1})})$$

with

$$g = \sqrt{(A^2 - 1)/\epsilon^2}$$

Note that N will in general not be an integer and must be rounded up to the next integer.

Example 3.13.   Design a Type-I Chebyshev filter having $R_P = 0.5$ dB, $A_S = 40$ dB, $\Omega_P = 0.5$ rad/s, and $\Omega_S = 0.65$ rad/s.

Using the above equations, the code is straightforward to write. Code has also been included to compute the realized or net values of R_P and A_S. Figure 3.12 shows the result.

```
function [Z,P,K] = LVDesignCheby1Filter(Rp,As,OmgP,OmgS)
% [Z,P,K] = LVDesignCheby1Filter(0.5,40,0.5,0.65)
E = sqrt(10^(Rp/10)-1); A = 10^(As/20); OmgC = OmgP;
OmgT = OmgS/OmgP; g = sqrt((A^2-1)/(E^2));
N = ceil(log10(g + sqrt(g^2 - 1))/log10(OmgT +...
```

```
    sqrt(OmgT^2-1)))
k = -(N-1):2:N-1; r = k*pi/(2*N); v0 = asinh(1/E)/N;
P = OmgC*(-sinh(v0)*cos(r) + j*cosh(v0)*sin(r)),
K = prod(abs(P)); Z = []; if rem(N,2)==0 % N even
K = 1/sqrt(1 + E^2)*K; end
```

The following code will obtain the Chebyshev poles and zeros, compute and plot the frequency response, and compute the realized values of R_P and A_S as well as the cascade coefficients:

```
Rp = 0.5; As = 40; OmgP = 0.5; OmgS = 0.65;
[Z,P,K] = LVDesignCheby1Filter(Rp,As,OmgP,OmgS)
H = LVsFreqResp(K*poly(Z),poly(P),2*OmgS,15);
[NetRp,NetAs] = LVsRealizedFiltParamLPF(H,OmgP,...
OmgS,2*OmgS)
[Bbq,Abq,Gain] = LVDirToCascadeClassIIR(poly(Z),poly(P),K)
```

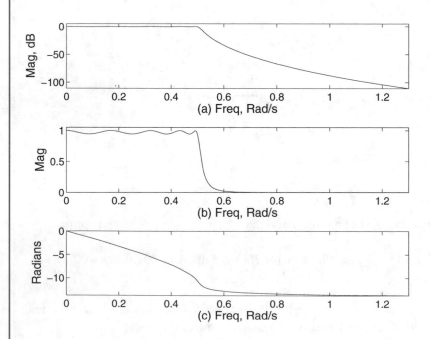

Figure 3.12: (a) Magnitude of response in dB of a Chebyshev Type-I filter designed to have $A_S = 40$ dB, $R_P = 0.5$ dB, $\Omega_P = 0.5$ rad/s, and $\Omega_S = 0.65$ rad/s; (b) Magnitude of response (linear) of same; (c) Phase response of same.

3.6   LOWPASS ANALOG CHEBYSHEV TYPE-II FILTERS

3.6.1   DESIGN BY ORDER, CUTOFF FREQUENCY, AND EPSILON

A Chebyshev Type-II filter is monotonic in the passband and equiripple in the stopband. One approach to deriving the magnitude squared characteristic of such a filter is to design a Chebyshev Type-I filter, substitute $1/\Omega$ for Ω, which converts the filter to highpass, and then subtract the result from 1 to convert it back to a lowpass filter.

We thus start with the magnitude squared response of a Type-I Chebyshev filter, which is

$$|H(\Omega)|^2 = \frac{1}{1 + \epsilon^2 (T_N(x))^2}$$

where $T_N(x)$ is the N-th order Chebyshev polynomial and $x = \Omega/\Omega_P$. We convert to a highpass filter by substituting $1/x$ for x

$$|H(\Omega)|^2 = \frac{1}{1 + \epsilon^2 (T_N(1/x))^2}$$

and then subtract from 1:

$$|H(\Omega)|^2 = 1 - \frac{1}{1 + \epsilon^2 (T_N(1/x))^2} = \frac{\epsilon^2 (T_N(1/x))^2}{1 + \epsilon^2 (T_N(1/x))^2}$$

Example 3.14.   Compute and plot the magnitude squared functions for Chebyshev Type-I and II filters having $\epsilon = 0.5$ and $N = 5$.

The following code follows the procedure outlined above. The result from running the code is shown in Fig. 3.13.

```
function LVCheby1toCheby2(Ep,N)
% LVCheby1toCheby2(0.5,5)
inc = 0.005; xLo = 0:inc:1;
xHi = 1+inc:inc:3; TnLo = cos(N*acos(xLo));
TnHi = cosh(N*acosh(xHi)); T = [TnLo,TnHi];
MagHSq = 1./(1 + Ep^2*(T.^2));
figure(33); subplot(211); xplot = [xLo, xHi];
plot(xplot,MagHSq); xlabel('Freq, Rad/s')
ylabel('Mag Squared')
% Convert to Type-II
xLo = xLo(find(~(xLo==0))); TnLo = cos(N*acos(1./xLo));
xHi = xHi(find(~(xHi==0))); TnHi = cosh(N*acosh(1./xHi));
T = [TnLo,TnHi]; MagHSq = 1 - 1./(1 + Ep^2*(T.^2));
subplot(212); xplot = [xLo, xHi];
```

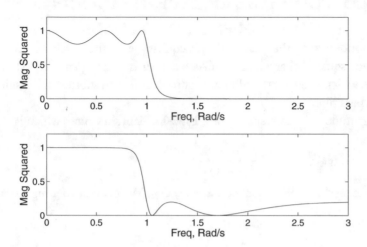

Figure 3.13: (a) Magnitude squared function for a Type-I Chebyshev filter having $\epsilon = 0.5$ and $N = 5$; (b) Magnitude squared function for a Type-II Chebyshev filter derived by substituting $1/\Omega$ for Ω and subtracting from 1, as described in the text.

> **plot(xplot,MagHSq); xlabel('Freq, Rad/s')**
> **ylabel('Mag Squared')**

MathScript provides a function to compute the poles and zeros of a normalized (i.e., $\Omega_C = 1.0$ rad/s) Chebyshev Type-II lowpass filter. The syntax is

$$[z, p, k] = cheb2ap(N, As)$$

where z is the vector of (finite) zeros, p is the vector of poles, k is the gain necessary to obtain unity magnitude response at $\Omega = 0$ rad/s, N is the desired order, and A_S is the stopband ripple (the minimum stopband attenuation) in dB.

Example 3.15.   Obtain the poles of a 3-rd order lowpass Chebyshev Type-II filter having $A_S = 40$ dB, scale them for $\Omega_C = 2$ rad/s, and compute the necessary value of k for the filter having $\Omega_C = 2$ rad/s. Write the system function. Plot the magnitude of frequency response.

We make the call

$$[z,p,k] = cheb2ap(3,40)$$

to obtain the normalized poles $p = [-0.1611 \pm j0.2959, -0.3523]$, which are then multiplied by the desired Ω_C (2 rad/s) to yield the poles for $\Omega_C = 2$ rad/s as $P = [-0.3222 \pm j0.5918, -0.7046]$. The zeros z are returned as $\pm j1.1547$, which when scaled by Ω_C yield $Z = [\pm j2.3094]$.

The net filter gain (at 0 rad/s) is the product of the magnitudes of the zeros divided by the product of the magnitudes of the poles, and the system function's numerator should be scaled by the reciprocal of this number. The following code uses the call above, scales the poles and zeros, computes K (the reciprocal of net filter gain), and computes and plots the frequency response from 0 rad/s up to twice the cutoff frequency Ω_C.

function LVCheb2(N,As,OmgC)
% LVCheb2(3,40,2)
[z,p,k] = cheb2ap(N,As)
z = OmgC*z, p = OmgC*p,
K = prod(abs(p))./prod(abs(z))
a = poly(p); b = K*poly(z);
H = LVsFreqResp(b,a,2*OmgC,16);
[Bbq,Abq,Gain] = LVDirToCascadeClassIIR(b/K,a,K)

The system function is

$$H(s) = \frac{(0.06)(s^2 + 5.333)}{(s^2 + 0.6446s + 0.4542)(s + 0.7046)}$$

3.6.2   DESIGN BY STANDARD PARAMETERS

Analogously to the procedure just described, Chebyshev Type-II filters are most easily derived by first computing the pole locations for a Type-I filter having the same specifications. The pole locations for the Type-I filter are the reciprocal of the pole locations for the Type-I filter. The Type-II also has finite zeros, which are located on the imaginary axis and computed as

$$\frac{j}{\sin(k\pi/2N)} \tag{3.19}$$

where

$$k = -(N-1):2:(N-1)$$

Note that when $\sin(k\pi/2N) = 0$, the frequency of the corresponding transfer function zero is infinite. Thus, some of the zeros for the Chebyshev Type-II filter may be infinite, while most will be finite.

The procedure to compute the reciprocal of the Type-I poles is often given as follows: if the Type-II poles being sought are represented as $\sigma_k' + j\Omega_k'$, and the Type-I poles as $\sigma_k + j\Omega_k$, then

$$\sigma_k' = \frac{\sigma_k}{\sigma_k^2 + \Omega_k^2}$$

and

$$\Omega_k' = \frac{\Omega_k}{\sigma_k^2 + \Omega_k^2}$$

The above procedure works since the set of poles comprise complex conjugate pairs; it actually computes the reciprocal of each pole's complex conjugate. Note that

$$1/(\sigma_k + j\Omega_k) = (\sigma_k - j\Omega_k)/(\sigma_k^2 + \Omega_k^2)$$

and

$$1/(\sigma_k - j\Omega_k) = (\sigma_k + j\Omega_k)/(\sigma_k^2 + \Omega_k^2)$$

A procedure to design a Chebyshev Type-II filter is as follows:

1) Specify the allowable passband ripple which is the minimum allowed response in the passband. This can be specified in positive dB, such as the typical parameter R_P.

2) Specify the maximum allowable response in the stopband. This can be done using A_S. Calculate ϵ from this value.

3) Calculate v_0 according to Eq. (3.17) and the poles for a Type-I filter according to Eq. (3.18).

4) Obtain the Type-II poles as the reciprocal of the Type-I poles.

5) Compute the Type-II zero locations according to Eq. (3.19).

Example 3.16.   Compute the poles and zeros of a Chebyshev Type-II filter having Ω_P = 0.9 rad/s, Ω_S = 1.0 rad/s, R_P = 0.2 dB, A_S = 40 dB.

The following code scales the desired frequency limits and computes a normalized Chebyshev Type-I filter. From this, the Type-II poles are obtained, the zeros separately computed, and then both poles and zeros are scaled to reflect the actual desired values of Ω_P and Ω_S. The result from running the code below is shown in Fig. 3.14.

```
function [Z,P,K] = LVDesignCheb2(Rp,As,OmgP,OmgS)
% [Z,P,K] = LVDesignCheb2(0.2,40,0.9,1)
OmgC = 1; OmgP = OmgP/OmgS;
E = 1/sqrt(10^(As/10)-1); G = 10^(-Rp/20);
N = acosh(G/(E*sqrt(1-G^2)))/acosh(1/OmgP);
N = ceil(abs(N)); V0 = asinh(1/E)/N;
k = -(N-1):2:(N-1); r = k*pi/(2*N);
Ch1P = -sinh(V0)*cos(r) + j*cosh(V0)*sin(r);
P = OmgS*(1./Ch1P); div = sin(k*pi/(2*N));
Zdiv = find(div==0); NZerDiv = div(find(~(div==0)));
Z = OmgS*(j./NZerDiv); s=j*[0:0.001:2*OmgS];
Hs = abs(polyval(poly(Z),s)./polyval(poly(P),s));
K = 1/max(abs(Hs));
```

The following code obtains the Chebyshev Type-II poles and zeros, computes and plots the frequency response, and computes the realized values of R_P and A_S as well as the cascade coefficients:

Rp = 0.2; As = 40; OmgP = 0.9; OmgS = 1;
[Z,P,K] = LVDesignCheb2(Rp,As,OmgP,OmgS)
H = LVsFreqResp(K*poly(Z),poly(P),2*OmgS,17);
[NetRp,NetAs] = LVsRealizedFiltParamLPF(H,OmgP,OmgS,2*OmgS)
[Bbq,Abq,Gain] = LVDirToCascadeClassIIR(poly(Z),poly(P),K)

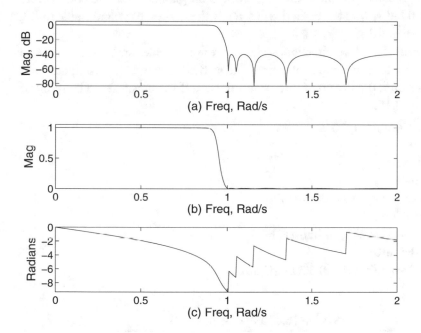

Figure 3.14: (a) Magnitude function in dB of a Type-II Chebyshev filter having $\Omega_P = 0.9$ rad/s, $\Omega_S = 1.0$ rad/s, $R_P = 0.2$ dB, and $A_S = 40$ dB; (b) Magnitude function (linear) of same; (c) Phase response of same.

3.7   ANALOG LOWPASS ELLIPTIC FILTERS

The Elliptic, or Cauer filter, has a steep roll-off with equiripple in both passband and stopband. Analogously to the equiripple FIR, it is possible to achieve the lowest order for a given set of specifications using an elliptic design.

The magnitude squared function of an Elliptic filter is

$$|H(j\Omega)|^2 = \frac{1}{1 + \epsilon^2 G_N^2(\Omega/\Omega_C)}$$

where ϵ is the passband ripple (related to R_P), and $G(\Omega/\Omega_C)$ is the N-th order Jacobian Elliptic function, the analysis of which is beyond the scope of this book. MathScript provides a function to compute the poles and zeros of a normalized (i.e., Ω_C = 1.0 rad/s) Elliptic lowpass filter. The syntax is

$$[z, p, k] = ellipap(N, Rp, As)$$

where z is the vector of (finite) zeros, p is the vector of poles, k is the gain necessary to obtain unity magnitude response at $\Omega = 0$ rad/s, N is the desired order, R_P is the desired passband ripple in dB, and A_S is the passband ripple in dB.

The following code calls the *ellipap* function, scales the resultant poles and zeros, computes the value K necessary to achieve a maximum of unity gain in the passband, and computes and plots the frequency response from 0 rad/s up to three times the cutoff frequency Ω_C. The result of computation is shown in Fig. 3.15.

```
function LVellip(N,Rp,As,OmgC)
% LVellip(5,0.2,40,2)
[z,p,k] = ellipap(N,Rp,As);
Z = OmgC*z, P = OmgC*p,
nrmGainFrZero = prod(abs(z))/prod(abs(p));
nrmK = k*nrmGainFrZero;
UnrmGnFrZero = prod(abs(Z))/prod(abs(P));
K = nrmK/UnrmGnFrZero;
H = LVsFreqResp(K*poly(Z),poly(P),3*OmgC,18);
```
The system function is

$$H(s) = \frac{0.1119(s^2 + 16.6703)(s^2 + 7.8158)}{(s^2 + 1.4092s + 2.8782)(s^2 + 0.3548s + 4.343)(s + 1.1663)}$$

3.7.1 DESIGN BY STANDARD PARAMETERS

The order N needed for an Elliptic filter to meet certain specifications can be computed from the following formula:

$$N = \frac{K(k)K(\sqrt{1 - k_1^2})}{K(k_1)K(\sqrt{1 - k^2})} \tag{3.20}$$

where

$$k = \frac{\Omega_P}{\Omega_S} \text{ and } k_1 = \frac{\epsilon}{\sqrt{A^2 - 1}} \tag{3.21}$$

and

$$K(x) = \int_0^{\pi/2} \frac{d\theta}{\sqrt{1 - x^2 \sin^2 \theta}} \tag{3.22}$$

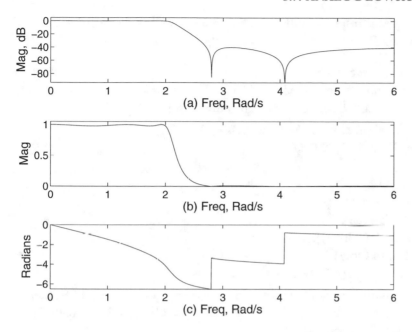

Figure 3.15: (a) Magnitude of frequency response in dB of an elliptic filter having $\Omega_C = 2$ rad/s, $N = 5$, $R_P = 0.2$ dB, $A_S = 40$ dB. (b) Magnitude of frequency response (linear) of same; (c) Phase response of same.

Eq. (3.22), which defines the **Complete Elliptic Integral of the First Kind** can be evaluated using the function

$$[k, e] = ellipke(m)$$

where k is the complete elliptic integral of the first kind, e is the complete elliptic integral of the second kind, and the modulus m would correspond to x^2 in Eq. (3.22).

Example 3.17.   Compute K(x) (the complete elliptic integral of the first kind) for x = 0.5.

We make the call

$$[k,e] = ellipke(0.5^2)$$

which yields $k = 1.6858$.

Example 3.18.   Write a script that receives the usual input specifications for a filter (R_P, A_S, W_P, W_S) and computes N, the poles, the zeros, and the gain for an elliptic filter meeting the specifications.

Compute and display the frequency response of the resulting filter as well as the realized values of R_P and A_S. Test the script using R_P = 1.25 dB, A_S = 50 dB, W_P = 0.5 rad/s, and W_S = 0.6 rad/s.

The code below computes the necessary order N from the specifications according to Eq. (3.20), and then completes the design by calling *LVellip*. The result from running the code is shown in Fig. 3.16 (the code for axis labels, etc., has been omitted for brevity).

```
function [Z,P,K] = LVDesignEllip(Rp,As,OmgP,OmgS)
% [Z,P,K] = LVDesignEllip(1.25,50,0.5,0.6)
E = (10^(Rp/10)-1)^0.5;
A=10^(As/20); OmgC = OmgP;
k = OmgP/OmgS; k1 = E/sqrt(A^2-1);
[K1k, K2k] = ellipke([k, (1-k^2)].^2);
[K1k1, K2k1] = ellipke([k1, (1-k1^2)].^2);
[Z,P,K] = LVellip(N,Rp,As,OmgC)
N = ceil(K1k(1)*K1k1(2)/(K1k1(1)*K1k(2)));
```

The following code obtains the Elliptic filter poles and zeros, computes and plots the frequency response, and computes the realized values of R_P and A_S as well as the Cascade Form coefficients:

```
Rp = 1.25; As = 50; OmgP = 0.5; OmgS = 0.6;
[Z,P,K] = LVDesignEllip(Rp,As,OmgP,OmgS)
H = LVsFreqResp(K*poly(Z),poly(P),3*OmgS,19);
[NetRp,NetAs] = LVsRealizedFiltParamLPF(H,OmgP,...
OmgS,3*OmgS)
[Bbq,Abq,Gain] = LVDirToCascadeClassIIR(poly(Z),poly(P),K)
```

3.8   FREQUENCY TRANSFORMATIONS IN THE ANALOG DOMAIN

It is possible to convert a prototype lowpass filter into a highpass, bandpass, and bandstop filter by substituting an expression in s for all values of s in the system function of the lowpass filter.

3.8.1   LOWPASS TO LOWPASS

To convert a prototype lowpass filter having passband cutoff Ω_P, replace each instance of s in the prototype lowpass system function with $s(\Omega_P / \Omega'_P)$, *i.e.*,

$$s \to s\frac{\Omega_P}{\Omega'_P}$$

Example 3.19.   Determine the system function for a Butterworth lowpass filter having Ω_C = 3 rad/s using the system function for a prototype Butterworth lowpass filter having Ω_C = 2 rad/s and N = 2.

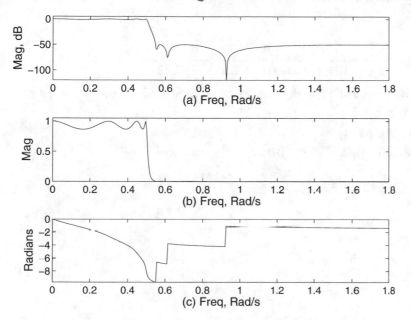

Figure 3.16: (a) Frequency response in dB of an elliptic filter having R_P = 1.25 dB, A_S = 50 dB, Ω_P = 0.5 rad/s, and Ω_S = 0.6 rad/s; (b) Frequency response (linear) of same; (c) Phase response of same.

To obtain the prototype lowpass filter we make the call

$$[z,p,k] = \text{buttap}(2)$$

to obtain the normalized poles p = [-0.7071 ± j0.7071], which are then multiplied by the desired Ω_C (2 rad/s) to yield the poles for Ω_C = 2 rad/s as P = [-1.414 ± j1.414]; from k, returned as 1.0, we obtain the new value of K as $k(\Omega_C^N) = (1.0)(2^2) = 4$. The system function for the prototype lowpass filter is therefore

$$H(s) = \frac{4}{(s + 1.414 + j1.414)(s + 1.414 - j1.414)}$$

To convert to a lowpass filter having Ω_C = 3 rad/s, we make the substitution

$$s \rightarrow \frac{\Omega_P}{\Omega'_P}s = \frac{2}{3}s$$

Thus, the new lowpass filter's system function is

$$H(s) = \frac{4}{(\frac{2}{3}s + 1.414 + j1.414)(\frac{2}{3}s + 1.414 - j1.414)}$$

which reduces to

$$H(s) = \frac{9}{(s + 2.121 + j2.121)(s + 2.121 - j2.121)}$$

The following code computes and plots the frequency response for both the prototype and new filter; the result is shown in Fig. 3.17.

```
[z,p,k] = buttap(2); p = 2*p; b1=4;
a1 = poly(p); p = [2.121*(-1 + j), 2.121*(-1 - j)];
b2=9; a2 = poly(p);
LVsFreqRespDouble(b1,a1,6,22,b2,a2)
```

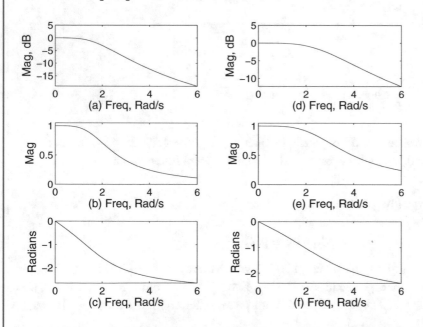

Figure 3.17: (a), (b), and (c): Magnitude (dB) response, Magnitude (linear) response, and phase response of prototype lowpass analog filter; respectively; (d), (e), (f): Magnitude (dB) response, Magnitude (linear) response, and phase response of new lowpass filter created from prototype lowpass analog filter; respectively.

The script

$$LVsFreqRespDouble(b1, a1, HiFreqLim, FigNo, b2, a2)$$

which was used in the code above, receives two sets of s-domain system function coefficients $b1, a1$, $b2, a2$, a high frequency limit for evaluation $HiFreqLim$, and a desired figure number $FigNo$ on which to plot the results:

```
function LVsFreqRespDouble(b1,a1,HiFreqLim,FigNo,b2,a2)
FR = (0:0.005:HiFreqLim); s = j*FR;
H = polyval(b1,s)./polyval(a1,s);
figure(FigNo); clf; subplot(321);
yplot = 20*log10(abs(H)+eps); plot(FR,yplot);
xlabel('(a) Freq, Rad/s'); ylabel('Mag, dB');
axis([0 inf -100 10]);
subplot(323); plot(FR,abs(H));
xlabel('(b) Freq, Rad/s'); ylabel('Mag');
axis([0 inf 0 1.1]);
subplot(325); plot(FR,unwrap(angle(H)))
xlabel('(c) Freq, Rad/s'); ylabel('Radians')
H = polyval(b2,s)./polyval(a2,s);
subplot(322); yplot = 20*log10(abs(H)+eps);
plot(FR,yplot); xlabel('(d) Freq, Rad/s');
ylabel('Mag, dB'); axis([0 inf -100 10]);
subplot(324); plot(FR,abs(H));
xlabel('(e) Freq, Rad/s'); ylabel('Mag');
axis([0 inf 0 1.1]); subplot(326);
plot(FR,unwrap(angle(H)))
xlabel('(f) Freq, Rad/s'); ylabel('Radians')
```

3.8.2   LOWPASS TO HIGHPASS

To convert a prototype lowpass filter having cutoff at Ω_P to a highpass filter having cutoff at Ω'_P, make the substitution

$$s \rightarrow \frac{\Omega_P \Omega'_P}{s}$$

Example 3.20.   Determine the system function for a Butterworth highpass filter having $\Omega_C = 3$ rad/s using the system function for a prototype Butterworth lowpass filter having $\Omega_C = 2$ rad/s and $N = 2$.

The system function for the prototype lowpass filter (as determined above and converting to coefficient form) is

$$H(s) = \frac{4}{(s^2 + 2.8284s + 4)}$$

To convert to a highpass filter having $\Omega_C = 3$ rad/s, we make the substitution

$$s \to \frac{\Omega_P \Omega'_P}{s} = \frac{(2)(3)}{s} = \frac{6}{s}$$

Thus, the new highpass filter's system function is

$$H(s) = \frac{4}{(36/s^2 + 2.8284(6/s) + 4)} = \frac{4s^2}{36 + 16.97s + 4s^2}$$

which reduces to

$$H(s) = \frac{s^2}{s^2 + 4.2426s + 9}$$

The following code computes and plots the frequency response for both the prototype and new filter: The result is shown in Fig. 3.18.

```
[z,p,k] = buttap(2); p = 2*p; b1=4;
a1 = poly(p); P = roots([1,4.2426,9]);
z = [0 0]; b2 = poly(z); a2 = poly(P);
LVsFreqRespDouble(b1,a1,6,25,b2,a2)
```

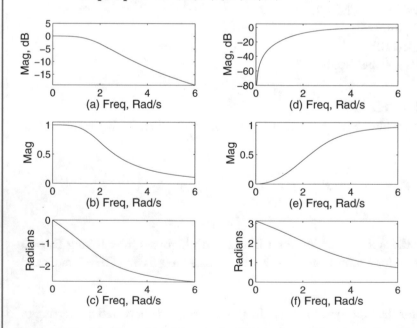

Figure 3.18: (a), (b), and (c): Magnitude (dB) response, Magnitude (linear) response, and phase response of prototype lowpass analog filter; respectively; (d), (e), (f): Magnitude (dB) response, Magnitude (linear) response, and phase response of new highpass filter created from prototype lowpass analog filter; respectively.

3.8.3 TRANSFORMATION VIA CONVOLUTION

At this point, we have seen how substitution of an expression in s for each instance of s in a system function can transform the system function into that of a different type of filter. Manual substitution, however, is time consuming and error-prone, so it is desirable, before proceeding to the even more complex substitutions for bandpass and bandstop filters, to develop a technique to automate the filter transformation.

The product of two polynomials can be obtained by convolution of the coefficients representing the polynomials. For example, if the polynomial

$$s^2 + 2s + 1$$

is represented as

$$[1, 2, 1]$$

then its square (for example), that is,

$$(s^2 + 2s + 1)^2$$

can be obtained as this convolution:

$$[1, 2, 1] \circledast [1, 2, 1] = [1, 4, 6, 4, 1] = s^4 + 4s^3 + 6s^2 + 4s + 1$$

where the symbol $\circledast$ here means linear convolution. Consider the generalized system function of an analog filter in factored form

$$H(s) = \Pi_{i=1}^{M}(s - z_m) / \Pi_{k=1}^{K}(s - p_k) \tag{3.23}$$

and consider a generalized form of a ratio of polynomials in s to be substituted for each instance of s in Eq. (3.23) as

$$\frac{b_2 s^2 + b_1 s + b_0}{a_2 s^2 + a_1 s + a_0}$$

which can be represented by coefficients only as

$$\frac{[b_2, b_1, b_0]}{[a_2, a_1, a_0]} = \frac{[N]}{[D]}$$

Each factor in Eq. (3.23), after making the substitution for s is of the form

$$(\frac{[N]}{[D]} - pz)$$

where pz represents a pole or zero of the system function.

Then the system function of the target filter $H(S)$ can be represented as the convolution of each factor represented in polynomial coefficient form

$$\{H(S)\} = \circledast_{i=1}^{M}(\frac{[N]}{[D]} - z_m)/ \circledast_{k=1}^{K} (\frac{[N]}{[D]} - p_k)$$

which reduces to

$$\{H(S)\} = \circledast_{i=1}^{M}(\frac{[N] - z_m[D]}{[D]})/ \circledast_{k=1}^{K} (\frac{[N] - p_k[D]}{[D]})$$

where $\{H(S)\}$ means "the polynomial coefficient representation of the target system function" and the symbols $\circledast_{i=1}^{M}$ and $\circledast_{k=1}^{K}$ mean the linear convolution of the factors in the numerator and denominator, respectively.

Several cases exist in the relationship of M to K, i.e., the number of zeros compared to the number of poles.

For Butterworth and Chebyshev Type-I filters, $M = 0$, that is, there are no finite zeros. Thus, the target system function in polynomial coefficient form is

$$\{H(S)\} = G/ \circledast_{k=1}^{K} (\frac{[N] - p_k[D]}{[D]})$$

which becomes

$$\{H(S)\} = \frac{G \circledast_{k=1}^{K} [D]}{\circledast_{k=1}^{K}([N] - p_k[D])} \tag{3.24}$$

where G is the gain of the prototype lowpass filter and

$$\circledast_{k=1}^{K}[D]$$

means the convolution of K factors $[D]$, such as, for example,

$$\circledast_{k=1}^{3}[D] = [D] \circledast [D] \circledast [D]$$

Chebyshev Type-II and Elliptic filters have zeros in their system functions. For N even, the number of zeros is equal to the number of poles ($M = K$), but for N odd, the number of zeros is one less than the number of poles.

When $M = K$, we get

$$\{H(S)\} = \frac{\circledast_{i=1}^{M}([N] - z_m[D])}{\circledast_{k=1}^{K}([N] - p_k[D])} \tag{3.25}$$

and when $M = K - 1$ we get

$$\{H(S)\} = \frac{[\circledast_{i=1}^{M}([N] - z_m[D])] \circledast [D]}{\circledast_{k=1}^{K}([N] - p_k[D])} \tag{3.26}$$

Example 3.21. Write a script that can transform a prototype Chebyshev Type-I lowpass filter of arbitrary order N into a highpass filter using polynomial convolution.

For a Chebyshev Type-I filter, $M = 0$ (there are no finite zeros), and for a highpass filter, the transformation is

$$s \to \frac{\Omega_P \Omega'_P}{s}$$

which can be represented in polynomial coefficient form as

$$\frac{[0, \Omega_P \Omega'_P]}{[1, 0]}$$

Each factor in the denominator of the lowpass filter's system function is

$$(s - p_k)$$

and each transformed factor can be represented as

$$(\frac{[N]}{[D]} - pk) = \frac{N - p_k[D]}{[D]} = \frac{[0, \Omega_P \Omega'_P] - p_k[1, 0]}{[1, 0]} \tag{3.27}$$

The following code first obtains the poles and gain (z is empty for a Chebyshev Type-I filter) for a normalized Chebyshev Type-I filter, scales p by Ω_P, computes the new lowpass filter gain G, establishes N and D in accordance with Eq. (3.27), convolves each of the pole factors, scales the coefficients so that the coefficient of the highest power in the denominator is 1, and finally computes and displays the frequency response. For purposes of testing, the values for N, $WpLP$ (the prototype lowpass filter Ω_P), $WpHP$ (the highpass filter Ω_P), and R_P have been chosen as 7, 3 rad/s, 3 rad/s, and 1 dB, respectively, but may be varied as desired. The result from running the code is shown in Fig. 3.19.

```
function [Z,P,K] = LVCheb1Lpf2Hpf(N,Rp,WpLP,WpHP)
% [Z,P,K] = LVCheb1Lpf2Hpf(7,1,3,3)
[z,p,k] = cheb1ap(N,Rp);
P = WpLP*p; G = k*WpLP^(length(P));
b1 = G; a1 = poly(P);
Num = 1; Den = 1; D = [1,0];
c = WpLP*WpHP; N = [0,c];
for Ctr = 1:1:length(P)
```

```
Num = conv(Num,N-P(Ctr)*D);
Den = conv(Den,D); end
a = real(Num); Scale = 1/a(1);
b2 = real(Den); a2 = Scale*a;
K = Scale*G; Z = roots(b2); P = roots(a2);
LVsFreqRespDouble(b1,a1,2*WpHP,29,K*b2,a2)
```

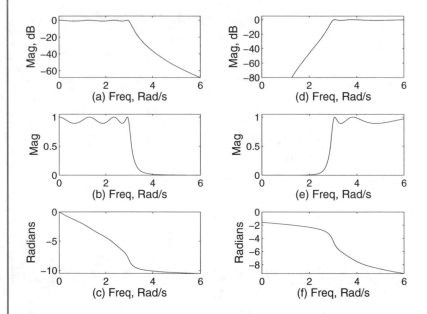

Figure 3.19: (a) Frequency response (dB) of a prototype Chebyshev Type-I lowpass filter having $\Omega_P = 2$ rad/s; (b) Frequency response (linear) of same; (c) Phase response of same; (d) Frequency response (dB) of a highpass filter having $\Omega_P = 6$ rad/s that was created by frequency domain transformation using the convolution method described in the text; (e) Frequency response (linear) of same; (f) Phase response of same.

3.8.4   LOWPASS TO BANDPASS

This method for transforming a prototype lowpass filter into a bandpass filter allows you to specify the new bandpass filter's four band edge frequencies, namely, Ω_{S1}, Ω_{P1}, Ω_{P2}, Ω_{S2}, and from these you compute the parameter Ω_0 (the "center" frequency) for the new bandpass filter

$$\Omega_0 = \sqrt{\Omega_{P1}\Omega_{P2}}$$

as well as the necessary values of Ω_P and Ω_S for the prototype lowpass filter:

$$\Omega_P = \frac{\Omega_{P2}^2 - \Omega_0^2}{\Omega_{P2}}$$

and

$$\Omega_S = \min \left\{ \frac{\Omega_{S2}^2 - \Omega_0^2}{\Omega_{S2}}, \frac{\Omega_0^2 - \Omega_{S1}^2}{\Omega_{S1}} \right\}$$

The prototype lowpass filter is then designed and in its system function, the following substitution is made:

$$s \to \frac{s^2 + \Omega_0^2}{s} \tag{3.28}$$

Example 3.22.   Design a Chebyshev Type-I bandpass filter that has the following band edges: Ω_{S1} = 4 rad/s, Ω_{P1} = 5 rad/s, Ω_{P2} = 8 rad/s, and Ω_{S2} = 10 rad/s. Use the method of convolution to obtain the new filter's system function.

The substitution shown in Eq. (3.28) may be represented as

$$\frac{N}{D} = \frac{[1,0,\Omega_0^2]}{[0,1,0]}$$

```
function [Z,P,K] = LVDesignCheby1BPF(OmS1,...
OmP1,OmP2,OmS2,Rp,As)
% [Z,P,K] = LVDesignCheby1BPF(4,5,8,10,1,40)
Om0Sq = OmP1*OmP2;
OmPlp = (OmP2^2 - Om0Sq)/OmP2;
OmSlp1 = (OmS2^2 - Om0Sq)/OmS2;
OmSlp2 = (Om0Sq - OmS1^2)/OmS1;
OmSlp = min([OmSlp1,OmSlp2]);
[Z1,P1,K1] = LVDesignCheby1Filter(Rp,As,OmPlp,OmSlp);
Num = 1; Den = 1; N = [1,0,Om0Sq]; D = [0,1,0];
for Ctr = 1:1:length(P1)
Num = conv(Num,[N-P1(Ctr)*D]);
Den = conv(Den,D); end; a = real(Num); b = real(Den);
s=j*[0:0.001:2*OmS2]; Hs = abs(polyval(b,s)./polyval(a,s));
K = 1/max(Hs); Z = roots(b); P = roots(a);
LVsFreqRespDouble(K1*poly(Z1),poly(P1),2*OmS2,30,K*b,a)
```

The following code, which uses the function above, obtains the new bandpass filter poles and zeros, computes and displays the frequency response of both the prototype lowpass filter and the new bandpass filter, computes the cascade coefficients, and computes the realized values of R_P and A_S for the design band edges using the script *LVxRealizedFiltParamBPF* (see exercises below).

> **OmS1=4; OmP1=5; OmP2=8; OmS2=10; Rp=1; As=40;**
> **[Z,P,K] = LVDesignCheby1BPF(OmS1,OmP1,OmP2,OmS2,Rp,As)**
> **[Bbq,Abq,Gain] = LVDirToCascadeClassIIR(poly(Z),poly(P),K)**
> **H = LVsFreqResp(K*poly(Z),poly(P),3*OmS2,19);**
> **[NetRp,NetAs1,NetAs2] = LVxRealizedFiltParamBPF(H,...**
> **OmS1,OmP1,OmP2,OmS2,3*OmS2)**

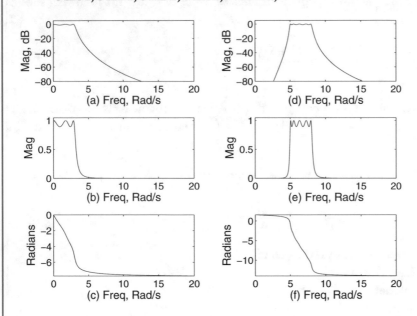

Figure 3.20: (a) Magnitude response (dB) of prototype lowpass filter having its Ω_P and Ω_S computed to result in a new bandpass filter meeting certain specifications after transformation via variable substitution implemented via convolution; (b) Magnitude response (linear) of same; (c) Phase response of same; (d) Magnitude response (dB) of the new bandpass filter; (e) Magnitude response (linear) of same; (f) Phase response of same.

3.8.5   LOWPASS TO BANDSTOP (NOTCH)

This method for transforming a prototype lowpass filter into a bandstop (or notch) filter allows you to specify the new notch filter's four band edge frequencies, namely, Ω_{P1}, Ω_{S1}, Ω_{S2}, Ω_{P2}, and from these you compute the parameter Ω_0 (the "center" frequency) for the new notch filter as

$$\Omega_0 = \sqrt{\Omega_{S1}\Omega_{S2}} \tag{3.29}$$

as well as the necessary values of Ω_P and Ω_S for the prototype lowpass filter:

$$\Omega_P = \frac{\Omega_{P1}}{\Omega_0^2 - \Omega_{P1}^2} \tag{3.30}$$

and

$$\Omega_S = \min\left\{\frac{\Omega_{S2}}{\Omega_{S2}^2 - \Omega_0^2}, \frac{\Omega_{S1}}{\Omega_0^2 - \Omega_{S1}^2}\right\} \tag{3.31}$$

The prototype lowpass filter is then designed and in its system function, the following substitution is made:

$$s \to \frac{s}{s^2 + \Omega_0^2} \tag{3.32}$$

Example 3.23.   Design a Chebyshev Type-I notch filter that has the following bandedges: Ω_{P1} = 4 rad/s, Ω_{S1} = 5 rad/s, Ω_{S2} = 8 rad/s, and Ω_{P2} = 10 rad/s. Use the method of convolution to obtain the new filter's system function. Compute and display the frequency responses of both prototype and notch filters.

The substitution shown in Eq. (3.32) may be represented as

$$\frac{N}{D} = \frac{[0,1,0]}{[1,0,\Omega_0^2]} \tag{3.33}$$

A straightforward implementation of Eqs. (3.29)–(3.32) and the convolution method using N and D as shown in Eq. (3.33) results in the following code; the bandstop filter frequency response obtained by making the call

[Z,P,K] = LVDesignCheby1Notch(4,5,8,10,1,40)

is shown in Fig. 3.21. The last code line of the function *LVDesignCheby1Notch* calls another new function, *LVxRealizedFiltParamNotch* (see exercises below), which, for a notch filter, determines the realized values of R_P and A_S for the design band edges $\Omega_{P1}, \Omega_{S1}, \Omega_{S2}, \Omega_{P2}$.

```
function [Z,P,K] = LVDesignCheby1Notch(OmP1,...
OmS1,OmS2,OmP2,Rp,As)
% [Z,P,K] = LVDesignCheby1Notch(4,5,8,10,1,40)
Om0Sq = OmP1*OmP2; OmPlp = OmP1/(Om0Sq-OmP1^2);
OmSlp1 = OmS2/(OmS2^2 - Om0Sq);
OmSlp2 = OmS1/(Om0Sq - OmS1^2);
OmSlp = min([OmSlp1,OmSlp2]);
[Z1,P1,K1] = LVDesignCheby1Filter(Rp,As,OmPlp,OmSlp);
H = LVsFreqResp(K1*poly(Z1),poly(P1),2*OmSlp,19);
```

```
Num = 1; Den = 1; D = [1,0,Om0Sq]; N = [0,1,0];
for Ctr = 1:1:length(P1)
Num = conv(Num,[N-P1(Ctr)*D]);
Den = conv(Den,D); end;
a = real(Num); b = real(Den); s = j*[0:0.001:2*OmS2];
Hs = abs(polyval(b,s)./polyval(a,s)); K = 1/max(Hs);
Z = roots(b); P = roots(a); H = LVsFreqResp(K*b,a,2*OmS2,31);
[NetAs,NetRp1,NetRp2] = LVxRealizedFiltParamNotch(H,OmP1,...
OmS1,OmS2,OmP2,2*OmS2)
```

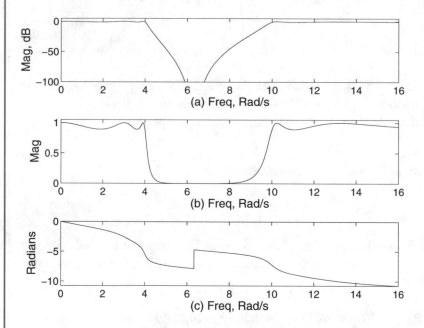

Figure 3.21: (a) Magnitude of response (dB) of bandstop (notch) filter designed by transforming a prototype Chebyshev Type-I lowpass filter according to the second method described in the text; (b) Magnitude of response (linear) of same; (c) Phase response of same.

3.9   ANALOG TO DIGITAL FILTER TRANSFORMATION

Having designed a filter in the analog domain, it is then necessary, to obtain a digital filter, to transform the poles and zeros into the z-domain in a way that preserves desirable attributes of the analog filter. Different transform methods exist to preserve different analog filter properties. The two most popular methods of doing this are the **Impulse Invariance Method** and the **Bilinear Transform**.

3.9.1   IMPULSE INVARIANCE

The Impulse Invariance Method attempts to preserve the analog filter's impulse response by sampling it at the time interval T. Since the digital filter's impulse response is a sampled version of the analog filter's impulse response, we can say that

$$H(z) = \sum_{n=0}^{\infty} h[n]z^{-n} \tag{3.34}$$

where

$$h[n] = h(nT)$$

where $h[n]$ is the sampled version of $h(t)$, the analog filter's impulse response. Since the classical IIR filters have system functions that are all ratios of polynomials in s, with the order of the numerator equal to or less than that of the denominator, a partial fraction expansion of $H(s) = B(s)/A(s)$ can be made:

$$H(s) = \sum_{j=1}^{N} \frac{R_j}{s + p_j} \tag{3.35}$$

The equivalent impulse response may be obtained then as

$$h(t) = \sum_{j=1}^{M} R_j e^{-p_j t}$$

and the sampled version is

$$h[n] = h[nT] = \sum_{j=1}^{M} R_j e^{-p_j nT} = \sum_{j=1}^{M} R_j (e^{-p_j T})^n$$

Substituting into Eq. (3.34) we get

$$H(z) = \sum_{n=0}^{\infty} \left[\sum_{j=1}^{M} R_j (e^{-p_j T})^n \right] z^{-n} \tag{3.36}$$

which results in

$$H(z) = \sum_{j=1}^{M} \frac{R_j z}{z - e^{-p_j T}} = \sum_{j=1}^{M} \frac{R_j}{1 - (e^{-p_j T})z^{-1}} \tag{3.37}$$

Since a sampling operation is involved, it follows that aliasing of the analog filter's frequency response must occur. The digital filter's transfer function is

$$H(z) = F_S \sum_{k=-\infty}^{\infty} H_a(s - j2\pi k F_S)$$

Thus, frequency strips of width $2\pi k F_S$ (= $2\pi k/T$) are folded or aliased into the frequency range $-\pi F_S$ to πF_S. Since no analog lowpass filter's transfer function is identically zero above any given finite frequency, it follows that degradation of the digital filter's transfer function relative to that of the analog filter must occur to a greater or lesser extent. This mapping of each horizontal strip of height $2\pi/T$ lying to the left of the s-plane imaginary axis into the interior of the unit circle in the z-plane is shown in Fig. 3.22.

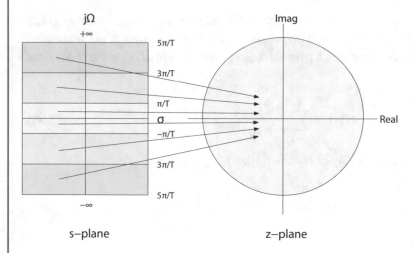

Figure 3.22: The Impulse Invariance mapping from the s-plane to the z-plane, due to aliasing, maps all horizontal strips of height $2\pi/T$ to the same place in the z-plane. The mapping is stable since values of s lying to the left of the s-plane imaginary axis map into the interior of the unit circle.

A procedure to perform the Impulse Invariance transformation is as follows:

- Set the desired digital filter values of ω_P and ω_S, pick a value for T_S (=$1/F_S$) and then compute the analog filter design frequencies Ω_P and Ω_S as

$$\Omega_P = \omega_P/T_S = \omega_P F_S$$

$$\Omega_S = \omega_S/T_S = \omega_S F_S$$

- Design a desired analog filter prototype using Ω_P and Ω_S, as well as desired values of R_P and A_S, and obtain its numerator and denominator system function coefficients, b and a.

- Perform a partial fraction expansion on b and a of the form given in Eq. (3.35).

- Convert analog poles into digital poles in accordance with Eq. (3.37).

The above steps are performed by the MathScript function

$$[Bz, Az] = impinvar(b, a, Fs)$$

which receives the analog filter b and a cocfficients and a desired sample rate $(= 1/T)$ and delivers the z-transform coefficients of the digital filter.

Example 3.24.   Devise a script that can receive desired digital filter cutoff frequencies ω_P and ω_S and a desired sample rate Fs and design an appropriate lowpass analog filter of any of the standard four types, and transform the analog filter into a digital filter, and compute and plot the frequency response of both filters.

```
function LVDesignDigFiltViaImpInv(Wp,Ws,Rp,As,FiltType,Fs)
% LVDesignDigFiltViaImpInv(0.4*pi,0.5*pi,1,40,1,1)
OmP = Wp*Fs; OmS = Ws*Fs;
if FiltType==1
[Z,P,K] = LVDesignButterworth(OmP,OmS,Rp,As);
elseif FiltType==2
[Z,P,K] = LVDesignCheby1Filter(Rp,As,OmP,OmS);
elseif FiltType==3
[Z,P,K] = LVDesignCheb2(Rp,As,OmP,OmS);
else
[Z,P,K] = LVDesignEllip(Rp,As,OmP,OmS);
end; b = K*poly(Z); a = poly(P);
[BZ,AZ] = impinvar(b,a,Fs);
LVsFRzFrLog(b,a,OmP,BZ,AZ,97)
```

We have introduced a new frequency response script, *LVsFRzFRLog*, as follows:

```
function LVsFRzFrLog(sB,sA,OmegaC,zB,zA,FigNo)
LnLm = 8*OmegaC;
Sargs = 0:0.01:LnLm;
Sargs = [Sargs]; s = j*Sargs;
Hs = polyval(sB,s)./polyval(sA,s);
Zargs = 0:0.01:pi; z = exp(j*Zargs);
Hz = polyval(zB,z)./polyval(zA,z);
figure(FigNo); subplot(211); plot(Sargs,20*log10(abs(Hs+eps)))
xlabel('(a) Freq, Radians/s'); ylabel('Magnitude, dB');
axis([0,LnLm,-100,5]); subplot(212);
```

```
ploty = 20*log10(abs(Hz)+eps);
plot(Zargs/pi,ploty); grid on;
xlabel('(b) Freq, Units of \pi'); ylabel('Magnitude, dB');
axis([10^(-2),inf,-100,(max(ploty)+10)])
```

Example 3.25.   Design a Butterworth digital filter having $\omega_P = 0.5\pi$, $\omega_S = 0.7\pi$, $R_P = 0.5$ dB, and $A_S = 40$ dB using the script written for the previous example. Repeat with $\omega_P = 0.5\pi$ and $\omega_S = 0.65\pi$.

Letting $F_S = 1$ we make the call

$$\text{LVDesignDigFiltViaImpInv(0.5*pi,0.7*pi,0.5,40,1,1)}$$

and observe the results in Fig. 3.23.

Making the call

$$\text{LVDesignDigFiltViaImpInv(0.5*pi,0.65*pi,0.5,40,1,1)}$$

we obtain the results shown in Fig. 3.24, which show a definite degradation in the stopband. However, this occurs at an attenuation level exceeding the required 40 dB.

Carrying the Butterworth experiment one step further, we further narrow the transition band with the following call

$$\text{LVDesignDigFiltViaImpInv(0.5*pi,0.62*pi,0.5,40,1,1)}$$

which results in a total degradation of the filter as shown in Fig. 3.25.

Example 3.26.   Design a Chebyshev Type-I digital filter having $\omega_P = 0.5\pi$, $\omega_S = 0.7\pi$, $R_P = 0.5$ dB, and $A_S = 40$ dB. Repeat the design using $\omega_P = 0.5\pi$, $\omega_S = 0.525\pi$, $R_P = 0.5$ dB, and $A_S = 40$ dB.

We make the following calls

$$\text{LVDesignDigFiltViaImpInv(0.5*pi,0.7*pi,0.5,40,2,1)}$$

$$\text{LVDesignDigFiltViaImpInv(0.5*pi,0.525*pi,0.5,40,2,1)}$$

which result in Figs. 3.26 and 3.27.

Example 3.27.   Design a Chebyshev Type-II digital filter having $\omega_P = 0.5\pi$, $\omega_S = 0.7\pi$, $R_P = 0.5$ dB, and $A_S = 40$ dB.

We make the call

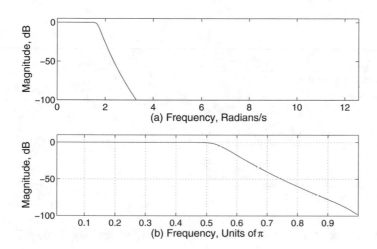

Figure 3.23: (a) Magnitude (dB) of response of prototype analog Butterworth filter; (b) Magnitude (dB) of response of digital Butterworth filter designed by Impulse Invariance.

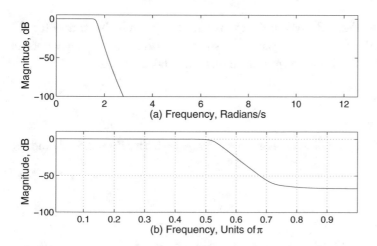

Figure 3.24: (a) Magnitude (dB) of response of prototype analog Butterworth filter; (b) Magnitude (dB) of response of digital Butterworth filter designed by Impulse Invariance.

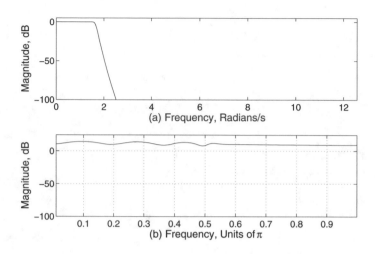

Figure 3.25: (a) Magnitude (dB) of response of prototype analog Butterworth filter; (b) Magnitude (dB) of response of digital Butterworth filter designed by Impulse Invariance.

LVDesignDigFiltViaImpInv(0.5*pi,0.7*pi,0.5,40,3,1)

the result from which is shown in Fig. 3.28. We note an unsatisfactory result. Note that for the Chebyshev Type-II analog filter, the frequency response does not go to zero as frequency goes to infinity, and the aliasing inherent in the Impulse Invariance technique takes its toll.

Example 3.28.   Design an Elliptic digital filter having $\omega_P = 0.5\pi$, $\omega_S = 0.7\pi$, $R_P = 0.5$ dB, and $A_S = 40$ dB.

We make the call

LVDesignDigFiltViaImpInv(0.5*pi,0.7*pi,0.5,40,4,1)

the result from which is shown in Fig. 3.29. Again we note an unsatisfactory result. Note that like the Chebyshev Type-II analog filter, the frequency response of an Elliptic analog filter does not go to zero as frequency goes to infinity, and once again the aliasing inherent in the Impulse Invariance technique leads to a poor digital filter.

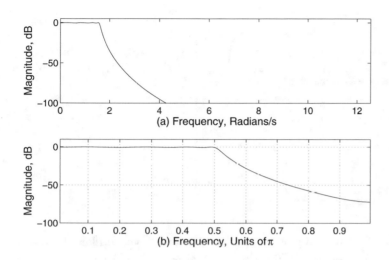

Figure 3.26: (a) Magnitude (dB) of response of prototype analog Chebyshev Type-I filter; (b) Magnitude (dB) of response of digital Chebyshev Type-I filter designed by Impulse Invariance.

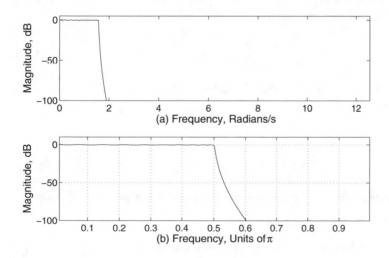

Figure 3.27: (a) Magnitude (dB) of response of prototype analog Chebyshev Type-I filter; (b) Magnitude (dB) of response of digital Chebyshev Type-I filter designed by Impulse Invariance.

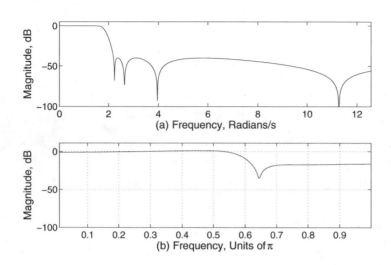

Figure 3.28: (a) Magnitude (dB) of response of prototype analog Chebyshev Type-II filter; (b) Magnitude (dB) of response of digital Chebyshev Type-II filter designed by Impulse Invariance.

3.9.2   THE BILINEAR TRANSFORM

A pole (or zero) in the Laplace domain can be mapped to a pole or zero in the z-domain using Eq. (3.38), which is known as the Bilinear transform.

$$z = \frac{1 + sT_s/2}{1 - sT_s/2} \tag{3.38}$$

where T_s is the sampling period of the digital system. To convert from a system function in the variable s, make the substitution

$$s = \frac{2}{T_s} \frac{1 - z^{-1}}{1 + z^{-1}}$$

Example 3.29.   Convert the Laplace domain system function $1/(s + 1)$ to the z-domain using T_s = 1.

This is a matter of algebraic substitution and fractional simplification:

$$H(z) = 1/(\frac{2}{T} \frac{(1 - z^{-1})}{(1 + z^{-1})} + 1) =$$
$$1/(\frac{2(1 - z^{-1}) + (1 + z^{-1})}{(1 + z^{-1})})$$

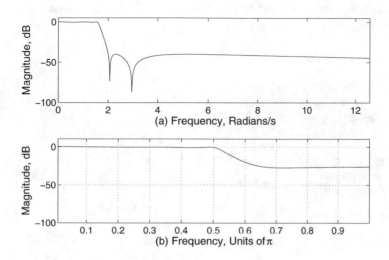

Figure 3.29: (a) Magnitude (dB) of response of prototype analog Elliptic filter; (b) Magnitude (dB) of response of digital Elliptic filter designed by Impulse Invariance.

which reduces to

$$H(z) = \frac{1 + z^{-1}}{3 - z^{-1}} = \frac{0.333(1 + z^{-1})}{1 - 0.333z^{-1}}$$

Note that the Laplace system function had one pole and one zero at infinite frequency. This has transformed into a z-domain system function having one pole and one zero at $z = -1$.

We can examine the Bilinear transform by letting $s = \sigma + j\omega$ and $T_s = 1/F_s$ in Eq. (3.38), which yields Eq. (3.39).

$$z = \frac{(2F_s + \sigma) + j\Omega}{(2F_s - \sigma) - j\Omega} \tag{3.39}$$

and the magnitude of z is

$$|z| = \frac{\sqrt{(2F_s + \sigma)^2 + \Omega^2}}{\sqrt{(2F_s - \sigma)^2 + \Omega^2}} \tag{3.40}$$

It can be seen from inspection of Eq. (3.40) that when $\sigma < 0$, Laplace poles are in the left half-plane (and are stable since $e^{\sigma t}e^{j\Omega t}$ decays to zero as $t \to \infty$ when $\sigma < 0$); these poles map to the interior of the unit circle i.e., $|z| < 1$) in the z-plane, which defines the stable region for z-plane poles. Similarly, Laplace poles on the imaginary axis (i.e., $\sigma = 0$), which generate constant, unity-amplitude time domain responses, map to the unit circle in the z-plane; z-plane poles lying on the unit circle also generate constant, unity-amplitude time domain responses. Figure 3.30 illustrates a few of the features of the Bilinear mapping.

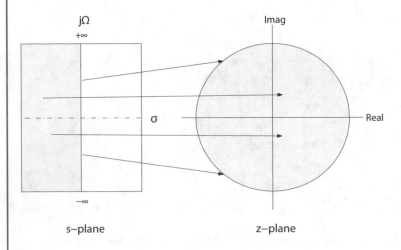

s–plane   z–plane

Figure 3.30: Mapping of the variable s in the Laplace (analog) plane to the variable z in the z- or digital domain. Note that the positive $j\Omega$ axis of the s-plane maps to the upper half of the unit circle in the z-plane, and the negative $j\Omega$ axis maps to the lower half of the unit circle. The upper (positive frequency) left half-plane of the Laplace domain maps to the upper interior of the unit circle, and the lower (negative frequency) half-plane maps to the lower interior of the unit circle. Both positive and negative infinity on the $j\Omega$ axis map to $z = -1$, and $s = 0 + j0$ maps to $z = +1$.

When a pole or pair of complex conjugate poles lies in the left half-plane, they map to the interior of the unit circle and both s- and z- impulse responses decay to zero. Figures 3.31 and 3.32 show the Laplace and z- domain poles and corresponding impulse responses for a single pole and a pair of complex conjugate poles, respectively.

When a pole lies on the imaginary axis in the s-domain (i.e., $\sigma = 0$), the impulse response is a constant, unity-amplitude sinusoid; this pole location is equivalent to a pole on the unit circle in the z-domain. This situation is depicted in Fig. 3.33. Poles in this location lead to marginally stable filters–for some signals, the filter will appear to be stable, while for other signals, it will be clearly unstable.

A pole to the right of the imaginary axis in the s-plane leads to an impulse response which is a complex exponential of continually increasing amplitude. This pole location corresponds to a pole outside the unit circle in the z-plane. Figure 3.34 depicts this situation.

System Transformation Via Convolution
The algebraic manipulations necessary to transform a Laplace system function into a z-domain system function are, in general, arduous and error-prone, so it behooves us to design a method for performing the operation by computer. The Laplace system function can be written as

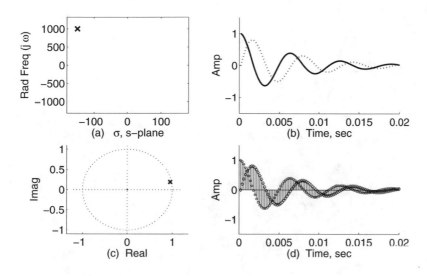

Figure 3.31: (a) A single pole in the left half-plane (i.e., $\sigma < 0$) in the Laplace Domain; (b) The real and imaginary parts of the impulse response corresponding to the pole plotted in (a); (c) A pole in the z-Plane, lying inside the Unit Circle, obtained using the Bilinear transform; (d) Real and imaginary parts of the impulse response corresponding to the pole in (c).

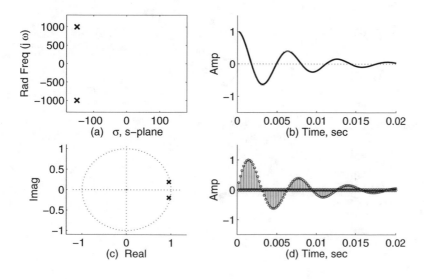

Figure 3.32: (a) A pair of complex conjugate poles in the left half-plane (i.e., $\sigma < 0$) in the Laplace Domain; (b) The (real-only) impulse response corresponding to the poles plotted in (a); (c) A pair of poles in the z-Plane, lying inside the Unit Circle, obtained using the Bilinear transform; (d) The (real-only) impulse response corresponding to the poles in (c).

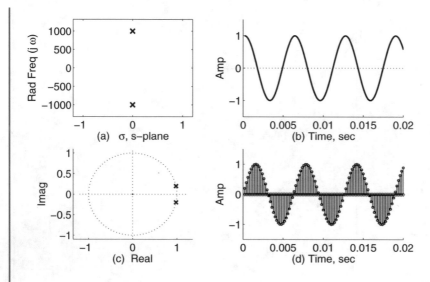

Figure 3.33: (a) A pair of complex conjugate poles on the Imaginary Axis (i.e., Damping = 0) in the Laplace Domain; (b) The real, undamped waveform generated by the poles plotted in (a); (c) The Bilinear-transformed pair of poles in the z-Plane, lying on the Unit Circle; (d) Real waveform in z-Plane generated by the poles in (c).

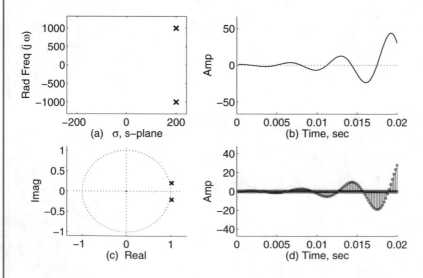

Figure 3.34: (a) A pair of complex conjugate poles to the right the Imaginary Axis (i.e., having gain greater than unity) in the Laplace Domain; (b) The real waveform generated by the poles plotted in (a); (c) The Bilinear-transformed pair of poles in the z-Plane, lying outside the Unit Circle; (d) Real waveform in z-Plane generated by the poles in (c).

$$H(s) = \Pi_{i=1}^{M}(s - z_m)/\Pi_{k=1}^{K}(s - p_k)$$

Making the substitution

$$s = \frac{2}{T_S}\frac{(1 - z^{-1})}{(1 + z^{-1})} = \frac{2F_S(1 - z^{-1})}{1 + z^{-1}}$$

the system function becomes

$$H(z) = \Pi_{i=1}^{M}\left(\frac{2F_S(1 - z^{-1})}{1 + z^{-1}} - z_m\right)/\Pi_{k=1}^{K}\left(\frac{2F_S(1 - z^{-1})}{1 + z^{-1}} - p_k\right) \qquad (3.41)$$

Each factor is of the form

$$\frac{2F_S(1 - z^{-1})}{1 + z^{-1}} - pz$$

where pz represents a pole or zero from the s-domain transfer function. We can represent each factor by equivalent coefficient arrays.

$$\frac{2F_S[1, -1]}{[1, 1]} - pz = \frac{2F_S[1, -1]}{[1, 1]} - \frac{pz[1, 1]}{[1, 1]}$$

which reduces to

$$\frac{[(2F_S - pz), -(2F_S + pz)]}{[1, 1]}$$

and the symbolic system function in the z-domain can be generated by the following ratio of convolutions:

$$\{H(z)\} = \circledast_{m=1}^{M}\left(\frac{[(2F_S - z_m), -(2F_S + z_m)]}{[1, 1]}\right)/ \circledast_{k=1}^{K}\left(\frac{[(2F_S - p_k), -(2F_S + p_k)]}{[1, 1]}\right) \qquad (3.42)$$

where z_m and p_k represent zeros and poles, respectively, of the s-domain transfer function. We will use this method in several examples in the following section.

Frequency Relationship & Pre-Warping

A pole in the Laplace domain, depending on its imaginary part, can generate frequencies from negative infinity to positive infinity; poles in the z-domain generate frequencies $-F_S/2$ to $F_S/2$. For frequencies that are at or below about F_S (the sampling rate used in the transform), the time domain response generated in the z-plane is similar to that generated in the Laplace plane, i.e., there is a close relationship between the Laplace and z-domain frequencies. As the Laplace frequency increases relative to F_S, the relationship becomes more and more nonlinear. Figure 3.35 depicts this relationship, with the Laplace frequency normalized to multiples of F_S.

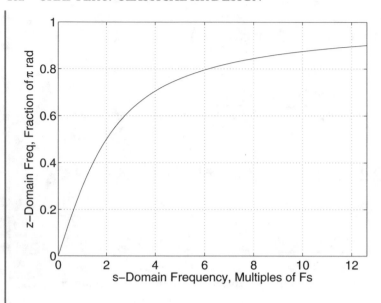

Figure 3.35: The highly nonlinear relationship between s-Domain frequency and z-Domain frequency.

As a result of the nonlinear frequency relationship between the s- and z- domains, it is necessary to pre-warp the s-domain poles so that after the Bilinear transform is performed they will be in desired locations in the z-domain.

The frequency relationships between the s- and z- domains are

$$\Omega = \frac{2}{T} \tan(\frac{\omega}{2}) \tag{3.43}$$

and

$$\omega = 2 \tan^{-1}(\frac{\Omega T}{2})$$

The following examples explore the concept of **Pre-Warping**, that is, choosing analog design frequencies that must be realized in the analog design so that after the Bilinear transform is performed, the digital domain frequencies are the ones specified in the digital design.

Example 3.30. Consider the single-pole analog filter having a 3 dB bandwidth of Ω and $H(s) = \Omega/(s + \Omega)$. It is desired to transform this system function into the z-domain with $\omega_c = 0.4\pi$ radians.

We solve for the value of Ω necessary to result, after the Bilinear transform is performed, in the desired ω_c.

$$\Omega = \frac{2}{T}\tan(\frac{0.4\pi}{2}) = \frac{2}{T}(0.7265) = \frac{1.45}{T}$$

and thus

$$H(s) = \frac{1.45/T}{s + 1.45/T}$$

From this, the analog filter design, we transform into the z-domain using the Bilinear transform:

$$H(z) = \frac{1.45/T}{2/T((1 - z^{-1})/(1 + z^{-1})) + 1.45/T} = \frac{1.45}{(2 - 2z^{-1})/(1 + z^{-1}) + 1.45}$$

which reduces to

$$H(z) = \frac{0.42(1 + z^{-1})}{1 - 0.1594z^{-1}}$$

Example 3.31.   Write a script suitable to convert the Laplace domain system function of a Butterworth or Chebyshev Type-I filter into a z-domain system function according to a given sample rate F_S. Test it by obtaining poles and zeros from the functions *buttap* and *cheb1ap* and converting them to the z-domain using a chosen value for Fs. Compute and plot the magnitude of frequency response of both the analog and digital filters.

Note that $\Omega_C = 1$ for the *buttap* and *cheb1ap* functions. You can enter your own value for Ω_C as *OmegaC* in the function *LVs2zViaBilinearEx* introduced below. With *OmegaC* = 1, try several values of F_S, such as 0.1, 0.5, 1, 2, 4, and 8, and plot ω_c(the z-domain cutoff frequency of the digital filter, observed on the plot resulting from running the code below) against the ratio of Ω_C to F_S. Since it is the ratio of a particular frequency in the analog domain, such as Ω_C, to F_S, rather than the actual values of these parameters, either of Ω_C or F_S may be arbitrarily chosen and held constant while the other is arbitrarily specified to achieve a given ω_c for the digital filter. After holding *OmegaC* at 1 and varying F_S, set $F_S = 1$ and vary Ω_C; you should be able to produce any desired ω_c for the digital filter using either method.

In general, necessary values for $\Omega_C, \Omega_P, \Omega_S$, etc., to achieve given frequencies in the z-domain while using a given (arbitrary) value for F_S can be computed using Eq. (3.43).

```
function LVs2zViaBilinearEx(N,Rp,OmegaC,Fs,CheborButter)
% LVs2zViaBilinearEx(3,1,1,1,1)
% LVs2zViaBilinearEx(3,1,1,5,1)
% LVs2zViaBilinearEx(3,1,5,5,1)
% LVs2zViaBilinearEx(3,1,5,25,1)
```

```
% LVs2zViaBilinearEx(4,[],4,8,0)
if CheborButter==1
[Z,P,K] = cheb1ap(N,Rp);
else; [Z,P,K] = buttap(N); end
K = K*OmegaC^N; P = OmegaC*P;
sB = K; sA = poly(P);
Num = 1; Den = 1; D = [1,1];
for Ctr = 1:1:length(P)
 N = [(2*Fs - P(Ctr)), -(2*Fs + P(Ctr))];
  Num = conv(Num,N); Den = conv(Den,D);
end; zB = real(Den); zA = real(Num);
LVsFRzFr(sB,sA,OmegaC,K*zB,zA,37)
```

In the script above, a new script was introduced which can receive a set of s-domain system coefficients sB and sA, a set of z-domain system coefficients zB and zA, a cutoff frequency $OmegaC$, and a desired figure number $FigNo$ to plot the frequency responses for both filters:

```
function LVsFRzFr(sB,sA,OmegaC,zB,zA,FigNo)
LnLm = 8*OmegaC;
rootrat = ((10^6)/(LnLm))^(0.001);
Sargs = 0:LnLm/1000:LnLm-LnLm/1000;
yy = (LnLm)*(rootrat.^[1:1:1000]);
Sargs = [Sargs yy]; s = j*Sargs;
Hs = polyval(sB,s)./polyval(sA,s);
Zargs = 0:0.01:pi; z = exp(j*Zargs);
Hz = polyval(zB,z)./polyval(zA,z);
figure(FigNo); subplot(211); semilogx(Sargs,abs(Hs))
xlabel('(a) Freq, Radians/s'); ylabel('Magnitude')
subplot(212); plot(Zargs/pi,abs(Hz))
xlabel('(b) Freq, Units of \pi'); ylabel('Magnitude')
```

The call

LVs2zViaBilinearEx(3,1,1,5,1)

for example, results in Fig. 3.36.

Example 3.32.   Write a script that can receive a design specification for a lowpass digital Chebyshev Type-I filter as R_P, A_S, ω_p, and ω_s, design an appropriate analog filter prototype, convert the prototype into a digital filter using the Bilinear transform implemented by convolution, and compute and display the frequency responses of both the analog prototype and digital filters.

The following script establishes F_S and then pre-warps the digital frequencies, i.e., computes the analog frequencies Ω_P and Ω_S (OmP and OmS in the script) necessary to result in the specified

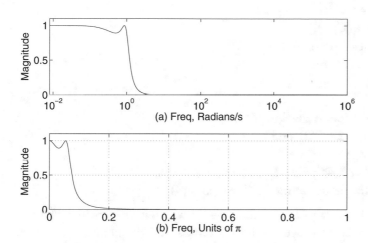

Figure 3.36: (a) A Chebyshev Type-I analog lowpass filter; (b) A digital version of the same, designed using the Bilinear Transform, with $\Omega_C = 1$ rad/s and $F_S = 1$. The lowpass cutoff frequency can be controlled in the digital domain by adjusting the ratio of Ω_C to F_S.

digital frequencies ω_p, and ω_s (*wp* and *ws* in the script) after the Bilinear transform is performed. The analog filter is then designed using R_P, A_S, Ω_P, and Ω_S, and then the analog frequency response is computed and displayed. The Bilinear transform is then computed, using a simplified convolution formula since the Chebyshev Type-I filter has no finite zeros (this script will also work with a Butterworth filter since it, too, lacks finite zeros in the analog domain). The digital filter's frequency response is then computed from its *b* and *a* coefficients (*SczNum* and *SczDen* in the script).

```
function LVsCheby1LPF2zCheby1LPF(Rp,As,wp,ws)
% LVsCheby1LPF2zCheby1LPF(1,40,0.5*pi,0.6*pi)
Fs = 1; T = 1/Fs; OmP = (2/T)*tan(wp/2);
OmS = (2/T)*tan(ws/2); E = sqrt(10^(Rp/10)-1);
A = 10^(As/20); OmC = OmP; OmT = OmS/OmP;
g = sqrt((A^2-1)/(E^2));
NN = ceil(log10(g + sqrt(g^2 - 1))/log10(OmT +...
 sqrt(OmT^2-1)))
v0 = asinh(1/E)/NN; k = -(NN-1):2:(NN-1);
P = OmC*(-sinh(v0)*cos(k*pi/(2*NN)) + ...
 j*cosh(v0)*sin(k*pi/(2*NN)));
NetK = prod(abs(P)); if rem(NN,2)==0
NetK = NetK/sqrt(1 + E^2); end
sB = NetK; sA = poly(P); Num = NetK; Den = 1;
```

```
for Ctr = 1:1:length(P); D = [1,1];
N = [(2*Fs-P(Ctr)), -(2*Fs+P(Ctr))];
Num = conv(Num,N); Den = conv(Den,D);
end; zB = real(Den); zA = real(Num);
b0 = 1/zB(1); a0 = 1/zA(1);
zA = zA*a0; zB = zB*b0;
G = abs(polyval(zA,exp(j*0))./polyval(zB,exp(j*0)));
if rem(NN,2)==0; G = G/sqrt(1 + E^2); end
zB = G*zB;
LVsFRzFr(sB,sA,OmC,zB,zA,44); Hz = LVzFr(zB,zA,1000,45);
[NetRp,NetAs] = LVsRealizedFiltParamLPF(Hz,wp,ws,pi)
```

The above script introduced the script *LVzFr*, the code for which is given below. This script allows you to specify a figure number on which to plot the results, and the number of frequency points to compute.

```
function Hz = LVzFr(zB,zA,NumPts,FigNo)
% Computes and displays on Figure FigNo the magnitude and
% phase response of a digital filter having numerator
% coefficients zB and denominator coefficients zA, for NumPts
% frequency points. The complex frequency response vector Hz is
% returned by the function and can be supplied to another script if
% desired to compute realized filter parameters.
Zargs = 0:pi/NumPts:pi; z = exp(j*Zargs);
Hz = polyval(zB,z)./polyval(zA,z);
figure(FigNo); subplot(211)
plot(Zargs/pi,20*log10(abs(Hz)+eps)); grid on
xlabel('(a) Freq, Units of \pi Radians');
ylabel('Magnitude, dB'); axis([0,1,-100,10])
subplot(212); plot(Zargs/pi,unwrap(angle(Hz))); grid on
xlabel('(b) Freq, Units of \pi Radians'); ylabel('Radians')
axis([0,1,-inf,inf])
```

3.10   MATHSCRIPT FILTER DESIGN FUNCTIONS

A number of functions exist in MathScript that design digital IIR filters using the Bilinear transform and analog prototype filters as described above. The following calls return the numerator and denominator coefficients for a lowpass digital filter having ω_C equal to the input argument OmC (supplied as a normalized frequency, $0 < \omega_C < 1.0$, where 1.0 is half the sampling rate), while N, R_P, and A_S have the usual meanings. In any of the calls given below, to obtain the output as zeros, poles, and gain, replace the output argument list $[b, a]$ with the list $[z, p, k]$.

[b,a] = butter(N,OmC)
[b,a] = cheby1(N,Rp,OmC)
[b,a] = cheby2(N,As,OmC)
[b,a] = ellip(N,Rp,As,OmC)

To design an analog filter, supply the trailing argument 's', and specify the desired Ω_C as any desired frequency in rad/s.

[b,a] = butter(N,OmC,'s')
[b,a] = cheby1(N,Rp,OmC,'s')
[b,a] = cheby2(N,As,OmC,'s')
[b,a] = ellip(N,Rp,As,OmC,'s')

A highpass digital filter can be designed with any one of the following calls; analog filters may be designed by supplying the trailing argument 's' and specifying the desired band limit in rad/s.

[b,a] = butter(N,OmC,'high')
[b,a] = cheby1(N,Rp,OmC,'high')
[b,a] = cheby2(N,As,OmC,'high')
[b,a] = ellip(N,Rp,As,OmC,'high')

By specifying two band limit frequencies for OmC, a digital bandpass filter can be designed with any one of the following calls, and analog bandpass filters may be designed by supplying the trailing argument 's' and specifying desired band limits in rad/s.

[b,a] = butter(N,[OmLo,OmHi])
[b,a] = cheby1(N,Rp,[OmLo,OmHi])
[b,a] = cheby2(N,As,[OmLo,OmHi])
[b,a] = ellip(N,Rp,As,[OmLo,OmHi])

Example 3.33.   Design an elliptic digital lowpass filter having $\Omega_L = 0.3\pi$ rad and $\Omega_H = 0.5\pi$ rad, $N = 7$, $R_P = 0.5$ dB, and $A_S = 50$ dB.

The following simple script designs the filter and plot its frequency response, which is shown in Fig. 3.37.

[b,a] = ellip(7,0.5,50,[0.3,0.5])
Hz = LVzFr(b,a,1000,21);

By specifying two band limit frequencies for OmC as well as the label 'stop', a bandstop or notch filter can be designed with any one of the following calls; analog filters may be designed using the same method described above for the preceding filter types.

[b,a] = butter(N,[OmLo,OmHi],'stop')
[b,a] = cheby1(N,Rp,[OmLo,OmHi],'stop')
[b,a] = cheby2(N,As,[OmLo,OmHi],'stop')
[b,a] = ellip(N,Rp,As,[OmLo,OmHi],'stop')

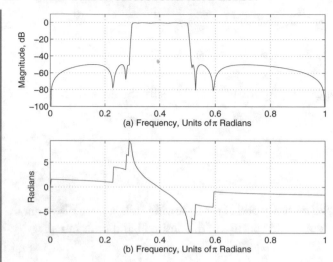

Figure 3.37: (a) Magnitude (dB) of an elliptic digital bandpass filter designed using the function *ellip*; (b) Phase response of same.

Example 3.34.   Design a Chebyshev Type-I digital notch filter having $\Omega_L = 0.3\pi$ rad and $\Omega_H = 0.5\pi$ rad, $N = 7$, $R_P = 0.25$ dB, and compute the realized value of R_P, and the four band edge frequencies for the realized value of R_P and a value of $A_S = 50$ dB.

The following simple script yields Fig. 3.38. Note that the MathScript design call allows you to specify the maximum value of R_P and the two passband edges; it does not allow you to specify the value of A_S in the design. The script *LVxBW4DigitalCheb1Notch* (see exercises below) locates the stopband edges for a user-specified value of A_S. The maximum value of ripple at the specified passband edges is also measured. The following numerical results were obtained using the script below: $\Omega_{S1} = 0.3369\pi$ rad, $\Omega_{S2} = 0.4568\pi$ rad for stopband attenuation of 50 dB, and the maximum value of ripple as $R_p = 0.25$ dB.

```
[b,a] = cheby1(7,0.25,[0.3,0.5],'stop')
Hz = LVzFr(b,a,1500,18);
[rS1,rS2,Rp1,Rp2] = LVxBW4DigitalCheb1Notch(Hz,0.3,0.5,50)
```

Example 3.35.   Design an elliptic digital bandstop filter having $\Omega_L = 3$ rad/s and $\Omega_H = 5$ rad/s, $N = 6$, $R_P = 0.25$ dB, and $A_S = 50$ dB. Obtain the output as poles, zeros, and gain.

The following simple script designs the filter and plot its frequency response, which is shown in Fig. 3.39.

```
[z,p,k] = ellip(6,0.25,50,[0.33,0.5],'stop')
Hz = LVzFr(k*poly(z),poly(p),1000,17);
```

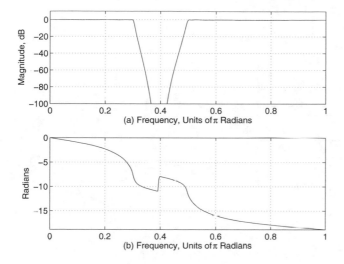

Figure 3.38: (a) Magnitude (dB) of a Chebyshev Type-I digital notch filter designed using the function *cheby*1; (b) Phase response of same.

3.11 PRONY'S METHOD

Usually, designing a filter entails starting with a desired frequency response, and computing either an impulse response in the case of an FIR, or poles and zeros in the case of an IIR.

Prony's method models an impulse response with a specified number of poles and zeros.

For the case of an IIR, if we have the impulse response $h[n]$ (or rather, a finite or truncated version of it), we can solve the IIR difference equation sequentially. The excitation sequence $x[n]$ is just the unit impulse, which is 1 for $n = 0$ and 0 otherwise. A general set of difference equations can be represented as

$$y[n] = \sum_{m=0}^{M} b_m x[n - m] - \sum_{p=1}^{N} a_p y[n - p]$$

and the first several equations generated thereby would be, for $N = 2$ and $M = 1$

$$y[0] = b_0 x[0] + b_1 x[-1] - a_1 y[-1] - a_2 y[-2]$$

$$y[1] = b_0 x[1] + b_1 x[0] - a_1 y[0] - a_2 y[-1]$$

$$y[2] = b_0 x[2] + b_1 x[1] - a_1 y[1] - a_2 y[0]$$

$$y[3] = b_0 x[3] + b_1 x[2] - a_1 y[2] - a_2 y[1]$$

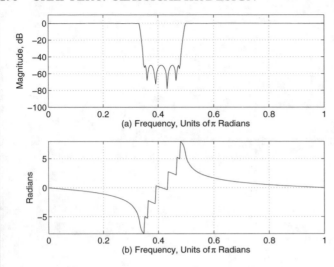

Figure 3.39: (a) Magnitude (dB) of an elliptic digital notch filter designed using the function *ellip*; (b) Phase response of same.

$$y[4] = b_0 x[4] + b_1 x[3] - a_1 y[3] - a_2 y[2]$$

Substituting in the known values of $x[n]$, and noting that $y[n] = h[n]$, the impulse response (since $x[n]$ is the unit impulse sequence), we get

$$h[0] = b_0 1 + b_1 0 - a_1 0 - a_2 0$$

$$h[1] = b_0 0 + b_1 1 - a_1 h[0] - a_2 0$$

$$h[2] = b_0 0 + b_1 0 - a_1 h[1] - a_2 h[0]$$

$$h[3] = b_0 0 + b_1 0 - a_1 h[2] - a_2 h[1]$$

$$h[4] = b_0 0 + b_1 0 - a_1 h[3] - a_2 h[2]$$

Moving all $h[n]$ terms to the left side, and writing the system in matrix form we get

$$Ha = b$$

where a and b are column vectors of the a and b coefficients; note that it is assumed that all coefficients were normalized by dividing by a_0, which is itself then equal to 1. The result, for $M = 2$ and $N = 1$, is

$$\begin{bmatrix} h[0] & 0 & 0 \\ h[1] & h[0] & 0 \\ h[2] & h[1] & h[0] \\ h[3] & h[2] & h[1] \end{bmatrix} \begin{bmatrix} 1 \\ a_1 \\ a_2 \end{bmatrix} = \begin{bmatrix} b_0 \\ b_1 \\ 0 \\ 0 \end{bmatrix}$$

This system can be exactly solved as given by noting that the lower two rows can be broken away and solved by themselves for the a coefficients; then the upper two rows can be solved for the b coefficients since the a coefficients would then be known. The lower system would be

$$\begin{bmatrix} h[2] & h[1] & h[0] \\ h[3] & h[2] & h[1] \end{bmatrix} \begin{bmatrix} 1 \\ a_1 \\ a_2 \end{bmatrix} = \begin{bmatrix} 0 \\ 0 \end{bmatrix}$$

We reconfigure the above into a solvable system by expanding into equations and reformatting into a matrix equation:

$$a_1 h[1] + a_2 h[0] = -h[2]$$

$$a_1 h[2] + a_2 h[1] = -h[3]$$

$$\begin{bmatrix} h[1] & h[0] \\ h[2] & h[1] \end{bmatrix} \begin{bmatrix} a_1 \\ a_2 \end{bmatrix} = \begin{bmatrix} -h[2] \\ -h[3] \end{bmatrix} \tag{3.44}$$

We get

$$H_{lp} a_{g0} = h_{lp}$$

where a_{g0} means the a coefficients with index greater than zero, h_{lp} means the impulse response, starting at an index one greater than the length of the desired b coefficient vector, and H_{lp} means the H matrix

$$\begin{bmatrix} h[0] & 0 & 0 \\ h[1] & h[0] & 0 \\ h[2] & h[1] & h[0] \\ h[3] & h[2] & h[1] \end{bmatrix}$$

starting with the second column and row index one greater than the number of desired b coefficients. Assuming that H_{lp} is nonsingular, we can obtain H_{lp}^{-1}, which is the pseudoinverse of H_{lp}, and solve for a_{g0} using standard matrix techniques:

$$H_{lp}^{-1} H_{lp} a_{go} = H_{lp}^{-1} h_{lp}$$

$$a_{go} = H_{lp}^{-1} h_{lp}$$

where a_{go} is returned as a column vector, and the net column vector of a coefficients is therefore

$$a = \begin{bmatrix} 1 \\ a_{go} \end{bmatrix}$$

The upper system of equations then allows us to solve the following matrix system for the b coefficients with a simple matrix multiplication:

$$\begin{bmatrix} h[0] & 0 & 0 \\ h[1] & h[0] & 0 \end{bmatrix} \begin{bmatrix} 1 \\ a_1 \\ a_2 \end{bmatrix} = \begin{bmatrix} b_0 \\ b_1 \end{bmatrix}$$

Consider the impulse response, the first few terms of which are

$$[1, 2.2, 2.05, 0.883, -0.5126]$$

and which we wish to model as an IIR having $a_0 = 1$, with $a_1, a_2, b_0,$ and b_1 to be determined. This impulse response was generated by an IIR having $b = [1, 0.9]$ and $a = [1, -1.3, 0.81]$. We get, from Eq. (3.44),

$$\begin{bmatrix} 2.2 & 1 \\ 2.05 & 2.2 \end{bmatrix} \begin{bmatrix} a_1 \\ a_2 \end{bmatrix} = \begin{bmatrix} -2.05 \\ -0.883 \end{bmatrix}$$

Setting this up in m-code, we get

Hlp = [2.2,1; 2.05,2.2]; hveclp = [-2.05,-0.883]'
a = pinv(Hlp)*hveclp
a = [1,a']'

which returns $a = [1, -1.3, 0.81]$. Solving for b, we get

Hup = [1,0,0;2.2,1,0];
b = Hup*a

which returns $b = [1, 0.9]$.

When a much longer portion of the impulse response is to be modeled, which is usually the case, we must resort to a least squares fit since the matrices involved are no longer square. Consider again the previous problem, but now we wish to model a long impulse response with the same few coefficients. The first few rows would be

$$\begin{bmatrix} h[0] & 0 & 0 \\ h[1] & h[0] & 0 \\ h[2] & h[1] & h[0] \\ h[3] & h[2] & h[1] \\ h[4] & h[3] & h[2] \\ \vdots & \vdots & \vdots \\ h[N] & h[N-1] & h[N-2] \end{bmatrix} \begin{bmatrix} 1 \\ a_1 \\ a_2 \end{bmatrix} = \begin{bmatrix} b_0 \\ b_1 \\ 0 \\ 0 \\ 0 \\ \vdots \\ 0 \end{bmatrix}$$

We must solve the lower set of equations, starting at row index one greater than the that of the b coefficients, using a least squares fit. Thus, notating the lower matrix system for this case as

$$H_{lp}a = b_{lp} \tag{3.45}$$

a least squares fit is defined as the set of a coefficients that minimize the sum of the squares of the differences between the elements of b_{lp} and $H_{lp}a$, i.e.,

$$\min \left| \sum (b_{lp} - H_{lp}a)^2 \right|$$

In order to obtain the values of a that obtain the minimum or least squared error case, we can solve Eq. (3.45) using the MathScript backslash operator. We illustrate this with an example.

Suppose we have the following set of equations in matrix form:

$$\begin{bmatrix} 1 & 1 & 1 \\ 1 & 2 & 4 \\ 1 & 3 & 9 \\ 1 & 4 & 16 \\ 1 & 5 & 25 \end{bmatrix} \begin{bmatrix} a_0 \\ a_1 \\ a_2 \end{bmatrix} = \begin{bmatrix} 1 \\ 4 \\ 9 \\ 16 \\ 25 \end{bmatrix} \tag{3.46}$$

which represent the algebraic system

$$y = a_0 x^0 + a_1 x^1 + a_2 x^2$$

for $x = [1{:}1{:}5]$. and $y = [1,4,9,16,25]$, i.e., the system of simultaneous equations

$$a_0 1^0 + a_1 1^1 + a_2 1^2 = 1$$

$$a_0 2^0 + a_1 2^1 + a_2 2^2 = 4$$

$$a_0 3^0 + a_1 3^1 + a_2 3^2 = 9$$

$$a_0 4^0 + a_1 4^1 + a_2 4^2 = 16$$

$$a_0 5^0 + a_1 5^1 + a_2 5^2 = 25$$

which can be written in matrix form as shown in Eq. (3.46) and written symbolically as

$$Ha = b$$

The system of equations above can be used to model the data defined by the x and y values with a second order polynomial, the coefficients of which are $[a_0, a_1, a_2]$. We can obtain a least squares-based estimate of a by making the MathScript call

$$a = b \backslash H$$

The values of a thus produced will minimize the squared error, i.e., we will obtain

$$\min \left| \sum (b - Ha)^2 \right|$$

Formulated in m-code, we have

h = [1,2,3,4,5]; b = [1,4,9,16,25]';
H = [h(1)^0, h(1)^1, h(1)^2; h(2)^0, h(2)^1, h(2)^2; ...
 h(3)^0, h(3)^1, h(3)^2; h(4)^0, h(4)^1, h(4)^2; ...
h(5)^0,h(5)^1,h(5)^2];
a = b\H

This yields the coefficient estimate as $[0.056, 0.23, 1]$. The actual coefficient values for this example are $[0,0,1]$. As the number of data points for this example increases, the coefficient estimate improves.

A simple program to construct a much larger matrix for the same problem might be as follows. Note that the variable *Lim* must be even for the indexing scheme to work.

Lim = 50; x = -Lim/2:1:Lim/2; for RowCtr = -Lim/2:1:Lim/2;
XX(RowCtr + Lim/2 + 1,1:1:3) = [1 x(RowCtr + Lim/2 + 1) ...
x(RowCtr+Lim/2+1)^2]; end;
Y = ([-Lim/2:1:Lim/2].^2)'; a = Y\XX

This yields a = $[0.0026,0,1]$, which is much closer to the ideal than when using just five data points. The reader will now be equipped to undertake, in the exercises below, the writing of the m-code for the script

$$LVxProny(Imp, NumA, NumB)$$

which receives an impulse response and a desired number of a coefficients $NumA$ and b coefficients $NumB$ to model the impulse response using a least squares fit.

Example 3.36.   Use Prony's method to find the numerator and denominator z-transform coefficients of an IIR that will produce the impulse response $(0.9)^n$, where n = 0:1:50.

We make the call

$$[b, a] = LVxProny([(0.9).\hat{}(0:1:50)],2,2)$$

which yields b = [1,0] and a = [1,-0.9].

MathScript also has a built-in function that performs Prony's Method, as follows:

$$[b, a] = prony(h, nNC, nDC)$$

where h is an impulse response to be modeled as an IIR with z-transform coefficients of b and a, nNC and nDC are the desired orders (one less than lengths) of the numerator and denominator of the z-transform, respectively.

3.12   IIR OPTIMIZATION PROGRAMS

In addition to IIR design using s-domain prototype filters, IIRs can be designed using programs that optimize one or more parameters to give a best fit to a given design specification. Reference [3] covers several basic methods to design IIR filters using frequency sampling. Reference [4] covers many different IIR optimization techniques in great detail.

3.13   REFERENCES

[1] Aram Budak, *Passive and Active Network Analysis and Synthesis*, Houghton Mifflin Company, Boston, Massachusetts, 1974.

[2] Richard G. Lyons, *Understanding Digital Signal Processing*, Addison Wesley Longman, Inc., Reading, Massachusetts, 1997.

[3] T. W. Parks and C. S. Burrus, *Digital Filter Design*, John Wiley & Sons, New York, 1987.

[4] Miroslav D. Lutovac, Dejan V. Tosic, and Brian L. Evans, *Filter Design For Signal Processing*, Prentice-Hall, Upper Saddle River, New Jersey 07458, 1996.

[5] William H. Beyer, editor, *CRC Standard Mathematical Tables, 26th Edition*, CRC Press, Inc., Boca Raton, Florida.

[6] John G. Proakis and Demitris G. Manolakis, *Digital Signal Processing, Principles, Algorithms, and Applications (Third Edition)*, Prentice Hall, Upper Saddle River, New Jersey 07458, 1996.

[7] Vinay K. Ingle and John G. Proakis, *Digital Signal Processing Using MATLAB V.4*, PWS Publishing Company, Boston, 1997.

3.14   EXERCISES

1. Compute the needed filter order for Butterworth, Chebyshev Type-I, Chebyshev Type-II, and Elliptic embodiments of an analog lowpass anti-aliasing filter. It must pass audio frequencies up to

20 kHz with no more than 1 dB of passband ripple, and the response at 22.05 kHz must be at least 60 dB below the peak passband response.

2. Compute the needed filter order for Butterworth, Chebyshev Type-I, Chebyshev Type-II, and Elliptic embodiments of an analog lowpass anti-aliasing filter for a system in which the sampling rate is 96 kHz; the passband ends at 20 kHz and must have no more than 1 dB or ripple, and the magnitude response at 48 kHz must fall to 60 dB below peak passband response.

3. Write a script that designs an analog Chebyshev Type-II bandpass filter according to the following function specification, and test it with the given test calls.

> **function [Z,P,K,NetRp,NetAs] = ...**
> **LVxDesignAnalogCheby2BPF(OmS1,OmP1,OmP2,OmS2,Rp,As)**
> **% Receives analog BPF band edges in radians/sec,**
> **% plus desired maximum Rp and minimum As and returns**
> **% the zeros (Z) and poles (P) of an analog Chebyshev**
> **% Type-II BPF along with filter gain K and the realized**
> **% values of Rp (NetRp) and As (NetAs).**
> **% Test calls:**
> **% [Z,P,K,NetRp,NetAs] = LVxDesignAnalogCheby2BPF(2,3,...**
> **% 6,8,1,45)**
> **% [Z,P,K,NetRp,NetAs] = LVxDesignAnalogCheby2BPF(1,...**
> **% 1.2,2.8,3.9,1,60)**
> **% [Z,P,K,NetRp,NetAs] = LVxDesignAnalogCheby2BPF(1,...**
> **% 1.2,4,5,1,75)**
> **% [Z,P,K,NetRp,NetAs] = LVxDesignAnalogCheby2BPF(4,...**
> **% 5,8,10,1,40)**
> **% [Z,P,K,NetRp,NetAs] = LVxDesignAnalogCheby2BPF(0.4*pi,...**
> **% 0.45*pi,0.6*pi,0.65*pi,0.5,50)**
> **% [Z,P,K,NetRp,NetAs] = LVxDesignAnalogCheby2BPF(19,...**
> **% 21,27,30,0.3,60)**

Use the lowpass-to-bandpass analog filter transformation method, implemented with the appropriate convolution transform method, and the Chebyshev Type-II prototype lowpass filter function *LVDesignCheb2*, all given in the text.

As part-and-parcel of this script, write the m-code for the script *LVxRealizedFiltParamBPF*, which is necessary to provide the realized values of R_P and A_S (considering both stopbands).

> **function [NetRp,NetAs1,NetAs2] = ...**
> **LVxRealizedFiltParamBPF(H,S1,P1,P2,S2,HiFreqLim)**
> **% Receives the four band edge design frequencies for a bandpass**
> **% filter, S1,P1,P2,and S2, the computed complex frequency**
> **% response H from frequency 0 up to HiFreqLim, and computes**
> **% the realized value of Rp (NetRp) for the passband and the**

% **realized values of As returns (NetAs1 and NetAs2) for each**
% **stopband.**

4. Write a script that designs an analog Elliptic bandstop filter according to the syntax below. Use the lowpass-to-bandstop analog filter transformation method, with the appropriate convolution transform method, and the Elliptic prototype lowpass filter function *LVDesignEllip*, all given in the text. The script should plot the magnitude and phase responses for the filter, and return the zeros, poles, and gain of the filter as well as the realized values of R_P and A_S. Test the script with the given test calls.

```
function [Z,P,K,NetRp,NetAs] = ...
LVxDesignAnalogEllipBPF(OmS1,OmP1,OmP2,OmS2,Rp,As)
% Receives analog BPF band edges in radians/sec, plus desired
% maximum Rp and minimum As and returns the zeros (Z) and
% poles (P) of an analog Elliptic BPF along with filter gain K and
% the realized values of Rp (NetRp) and As (NetAs), both in
% positive dB. The magnitude (in dB) and phase responses are
% also plotted.
% Test calls:
% [Z,P,K,NetRp,NetAs] = LVxDesignAnalogEllipBPF(2,3,...
% 6,8,1,45)
% [Z,P,K,NetRp,NetAs] = LVxDesignAnalogEllipBPF(1,1.2,...
% 2.8,3.9,1,60)
% [Z,P,K,NetRp,NetAs] = LVxDesignAnalogEllipBPF(1,1.2,...
% 4,5,1,75)
% [Z,P,K,NetRp,NetAs] = LVxDesignAnalogEllipBPF(4,5,...
% 8,10,1,40)
```

5. Write a script that can receive numerator and denominator coefficients of a classical IIR filter and a sampling rate and perform the Bilinear transform using the convolution method, returning the numerator and denominator coefficients of a digital filter. The script should conform to the following call; the script will be tested by using it to design digital filters in several following problems. The script should be able to deal with classical analog IIR filters, i.e., filters having either numerator and denominator coefficients of equal order, or having numerator coefficients of order one less than the denominator.

```
function [b,a] = LVxBilinearZPK(Z,P,K,Fs)
% Receives the zeros Z, poles P, and gain K of a classical IIR
% filter and performs the Bilinear transform using Fs as the
% sampling rate.
```

6. A digital signal sampled at 96 kHz will be digitally lowpass filtered to have its upper passband limit at 20 kHz, with no more than 1 dB or ripple, and the response at 22.05 kHz must have fallen by 60 dB below peak passband response. Compute the needed filter order for Butterworth, Chebyshev Type-I,

Chebyshev Type-II embodiments for an analog lowpass filter for this latter filtering operation, and then pick the lowest order filter type, pre-warp the digital band limits to obtain design band limits for an analog prototype filter, design the prototype filter and then convert it into a digital filter using the Bilinear transform as implemented by the function *LVxBilinearZPK* (see previous exercise). Compute the realized values of R_P and A_S for the digital filter.

7. Write a script that designs a digital elliptic bandpass filter according to the following syntax:

> **function [b,a,G,NetRp,NetAs] = LVxDesignDigEllipBPF(Rp,As,...**
> **ws1,wp1,wp2,ws2)**
> **% Receives digital BPF band edges in normalized frequency**
> **% (units of pi), plus desired maximum Rp and minimum As**
> **% and returns the b (numerator) and a (denominator)**
> **% coefficients of a digital Elliptic BPF along with filter gain G**
> **% and the realized values of Rp (NetRp) and As (NetAs). The**
> **% magnitude (in dB) and phase responses are also plotted.**
> **% Sample calls:**
> **% [b,a,G,NetRp,NetAs] = LVxDesignDigEllipBPF(1,45,...**
> **% 0.4,0.475,0.65,0.775)**
> **% [b,a,G,NetRp,NetAs] = LVxDesignDigEllipBPF(1,60,...**
> **% 0.45,0.55,0.7,0.83)**
> **% [b,a,G,NetRp,NetAs] = LVxDesignDigEllipBPF(1,75,...**
> **% 0.4,0.475,0.65,0.775)**
> **% [b,a,G,NetRp,NetAs] = LVxDesignDigEllipBPF(1,75,...**
> **% 0.1,0.12,0.4,0.5)**

Prewarp the digital filter specifications to obtain the analog prototype frequencies, and use the previously script *LVxDesignAnalogEllipBPF* for the previous exercise to design the analog protype. Use the Bilinear transform (implemented using the function *LVxBilinearZPK*) to obtain the digital filter coefficients from the analog prototype filter. Test your script with the given sample calls, and plot the magnitude and phase responses.

8. Modify the scripts *LVxDesignAnalogEllipBPF* and *LVxDesignDigEllipBPF* previously written into the following two scripts that perform the same functions for Chebyshev Type-II analog and digital filters. Test the scripts using the given test calls and plot the magnitude and phase responses of the designed filters.

> **function [Z,P,K,NetRp,NetAs] = ...**
> **LVxDesignAnalogCheby2BPF(OmS1,OmP1,OmP2,OmS2,Rp,As)**
> **% Receives analog BPF band edges in radians/sec, plus desired**
> **% maximum Rp and minimum As and returns the zeros (Z) and**
> **% poles (P) of an analog Chebyshev Type-II BPF along with filter**
> **% gain K and the realized values of Rp (NetRp) and As (NetAs).**
> **% Also plots the magnitude (in dB) and phase responses of both**

% the bandpass filter and lowpass prototype filter used in
% the design.
% Test calls:
% [Z,P,K,NetRp,NetAs] = LVxDesignAnalogCheby2BPF(2,3,...
% 6,8,1,45)
% [Z,P,K,NetRp,NetAs] = LVxDesignAnalogCheby2BPF(1,1.2,...
% 2.8,3.9,1,60)
% [Z,P,K,NetRp,NetAs] = LVxDesignAnalogCheby2BPF(1,1.2,...
% 4,5,1,75)
% [Z,P,K,NetRp,NetAs] = LVxDesignAnalogCheby2BPF(4,5,...
% 8,10,1,40)

function [b,a,G,NetRp,NetAs] = LVxDesignDigCheby2BPF(Rp,...
As,ws1,wp1,wp2,ws2)
% Receives digital BPF band edges in normalized frequency (units
% of pi), plus desired maximum Rp and minimum As and returns
% the b (numerator) and a (denominator) coefficients of a digital
% Cheby2 BPF along with filter gain G and the realized values
% of Rp (NetRp) and As (NetAs). Also plots the magnitude (in
% dB) and phase responses of the digital bandpass filter.
% Sample calls:
% [b,a,G,NetRp,NetAs] = LVxDesignDigCheby2BPF(1,45,...
% 0.4,0.475,0.65,0.775)
% [b,a,G,NetRp,NetAs] = LVxDesignDigCheby2BPF(1,60,...
% 0.45,0.55,0.7,0.83)
% [b,a,G,NetRp,NetAs] = LVxDesignDigCheby2BPF(1,75,...
% 0.4,0.475,0.65,0.775)
% [b,a,G,NetRp,NetAs] = LVxDesignDigCheby2BPF(1,75,...
% 0.1,0.12,0.4,0.5)

9. Write a script, as specified below, that designs an analog Chebyshev Type-II notch filter, and test the script with the given sample calls. Use the convolution method to derive the notch system function from an analog prototype lowpass Chebyshev Type-II filter.

function [Z,P,K,NetRp,NetAs] = ...
LVxDesignAnalogCheby2Notch(OmP1,OmS1,OmS2,OmP2,Rp,As)
% Receives analog Notch band edges in radians/sec,
% plus desired maximum Rp and minimum As and returns the
% zeros (Z) and poles (P) of an analog Chebyshev Type-II
% Notch along with filter gain K and the realized values of
% Rp (NetRp) and As (NetAs). Also plots the magnitude
% (in dB) and phase responses of both the notch filter and its

```
% lowpass analog prototype filter.
% Sample calls:
% [Z,P,K,NetRp,NetAs] = LVxDesignAnalogCheby2Notch(2,...
% 3,6,8,1,45)
% [Z,P,K,NetRp,NetAs] = LVxDesignAnalogCheby2Notch(1,...
% 1.2,2.8,3.9,1,60)
% [Z,P,K,NetRp,NetAs] = LVxDesignAnalogCheby2Notch(1,...
% 1.2,4,5,1,75)
% [Z,P,K,NetRp,NetAs] = LVxDesignAnalogCheby2Notch(4,...
% 5,8,10,1,40)
```

As part and parcel of writing the script *LVxDesignAnalogCheby2Notch*, write the m-code for the following script, which is necessary to compute the realized values of R_P and A_S for a notch filter:

```
function [NetAs,NetRp1,NetRp2] = ...
LVxRealizedFiltParamNotch(H,P1,S1,S2,P2,HiFreqLim)
% Receives the four band edge design frequencies for a notch
% filter, P1,S1,S2,and P2, the computed complex frequency
% response H from frequency 0 up to HiFreqLim, and returns the
% realized values of Rp (NetRp1 and NetRp2) for each passband
% and the realized value of As (NetAs) for the stopband.
```

10. Write a script, as specified below, that designs a digital Chebyshev Type-II notch filter. Design the prototype analog notch filter by prewarping the digital frequencies and calling the function written for the previous exercise (*LVxDesignAnalogCheby2Notch*), then use the Bilinear method (implemented using *LVxBilinearZPK*) to obtain the digital filter coefficients.

```
function [b,a,G,NetRp,NetAs] = ...
LVxDesignDigCheby2Notch(Rp,As,wp1,ws1,ws2,wp2)
% Receives digital Notch filter band edges in normalized frequency...
%  (units of pi), plus desired maximum Rp and minimum As and
% returns the b (numerator) and a (denominator) coefficients
% of a digital Cheby2 Notch filter along with filter gain G and
% the realized values of Rp (NetRp) and As (NetAs). The
% magnitude (in dB) and phase responses of both the notch
% filter and its analog prototype filter are also plotted.
% Sample calls:
% [b,a,G,NetRp,NetAs] = LVxDesignDigCheby2Notch(1,45,...
% 0.4,0.475,0.65,0.775)
% [b,a,G,NetRp,NetAs] = LVxDesignDigCheby2Notch(1,60,...
% 0.45,0.55,0.7,0.83)
% [b,a,G,NetRp,NetAs] = LVxDesignDigCheby2Notch(1,75,...
```

% 0.4,0.475,0.65,0.775)
% [b,a,G,NetRp,NetAs] = LVxDesignDigCheby2Notch(1,75,...
% 0.1,0.12,0.4,0.5)

11. Write a script that designs a Butterworth lowpass digital filter meeting the following specifications: ω_P, ω_S, R_P, and A_S in accordance with the call syntax below. Use the Bilinear transform method, implemented by the script *LVxBilinearZPK*, on a prototype analog lowpass Butterworth filter designed by prewarping the digital frequencies to obtain the necessary analog prototype frequencies. Plot the magnitude and phase responses of the digital filter, and return the output arguments as shown in the call syntax. Test the script with the given sample calls.

function [zB,zA,G,NetRp,NetAs] = ...
LVxDesignDigitalButterLPF(Rp,As,wp,ws)
% function [zB,zA,G,NetRp,NetAs] = ...
% LVxDesignDigitalButterLPF(Rp,As,wp,ws)
% Designs a digital Butterworth LPF having no more than Rp
% dB of ripple at passband edge wp, and at least As dB
% attenuation at stopband edge ws. The filter coefficients are
% returned as numerator coefficients zB, denominator
% coefficients zA, and gain G. The script also plots the
% magnitude (in dB) and phase responses of both the
% digital filter, and the analog prototype used to compute
% the digital filter coefficients using the Bilinear transform.
% The realized values of Rp and As are also returned.
% Sample calls:
% [zB,zA,G,NetRp,NetAs] = LVxDesignDigitalButterLPF(1,...
% 40,0.5*pi,0.6*pi)
% [zB,zA,G,NetRp,NetAs] = LVxDesignDigitalButterLPF(0.3,...
% 60,0.25*pi,0.325*pi)
% [zB,zA,G,NetRp,NetAs] = LVxDesignDigitalButterLPF(0.1,...
% 55,0.25*pi,0.35*pi)

12. Write the m-code for the script *LVxBW4DigitalCheb1Notch* as specified below, and test it with the given calls.

function [rS1,rS2,Rp1,Rp2] = LVxBW4DigitalCheb1Notch(H,...
OmP1,OmP2,As)
% H is the complex frequency response vector (from frequency 0
% to pi radians) of a Chebyshev Type I digital notch filter
% designed using the function cheby1; OmP1 and OmP2 are the
% passband edges, in normalized frequency (i.e., in multiples of
% pi radians) of a notch filter designed by the function cheby1,
% rS1 and rS2 are stopband edges at which the desired value of

% stopband attenuation As is realized. Rp1 and Rp2 are the
% realized or actual frequency responses (in positive dB) at the
% passband edges OmP1 and OmP2.

Test calls:

(a)

[b,a] = cheby1(6,0.35,[0.45, 0.7],'stop')

Hz = LVzFr(b,a,2000,29);

[rS1,rS2,Rp1,Rp2] = LVxBW4DigitalCheb1Notch(Hz,0.45,0.7,60)

(b)

[b,a] = cheby1(5,0.15,[0.25, 0.75],'stop')

Hz = LVzFr(b,a,2000,29);

[rS1,rS2,Rp1,Rp2] = LVxBW4DigitalCheb1Notch(Hz,0.25,0.75,55)

13. Write a script that conforms to the following syntax:

function LVxDigCheby1LPFViaMS(Rp,As,wp,ws)
% Performs the same function as
% LVsCheby1LPF2zCheby1LPF(Rp,As,wp,ws) but in addition it
% uses the MathScript function [b,a]=cheby1(N,Rp,wp) to compute
% the digital filter coefficients using the computed value for analog
% filter order and the desired passband edge wp. The magnitude
% (in dB) and phases responses of the digital filter computed both
% ways are displayed, and the realized filter parameters are
% computed for comparison.
% LVxDigCheby1LPFViaMS(1,40,0.5*pi,0.6*pi)
% LVxDigCheby1LPFViaMS(1,60,0.45*pi,0.65*pi)
% LVxDigCheby1LPFViaMS(1,68,0.5*pi,0.55*pi)

You should find that the results from both digital filter computation methods are essentially identical for the calls given above.

14. Write a script that compares the performance of the script

$$LVxDesignDigEllipBPF(Rp, As, ws1, wp1, wp2, ws2)$$

to the function *ellip*. Use one-half the order of the digital elliptic bandpass filter designed by *LVxDesignDigEllipBPF* as the order for the call to *ellip*, and use $wp1$ and $wp2$ (the passband edges) for the needed frequency vector. Plot the magnitude (dB) and phase responses of the filters designed by both scripts, and compute the realized values of R_P and A_S for both filters. You should be able to produce nearly identical results.

Your script to perform the above comparison should conform to the following syntax:

function LVxDigEllipBPFViaMS(Rp,As,ws1,wp1,wp2,ws2)
% Receives the values for maximum passband ripple Rp, minimum

% stopband attenuation As, and bandpass filter band edges ws1,
% wp1,wp2, and ws2, and designs digital bandpass filters using 1)
% the script LVxDesignDigEllipBPF and 2) the built-in function
% ellip. Plots the magnitude and phase responses of both designed
% filters as well as the realized values for Rp and As for both
% filters, allowing for visual and numerical comparison of the
% two designs.
% Test calls:
% LVxDigEllipBPFViaMS(1,75,0.4,0.475,0.65,0.775)
% LVxDigEllipBPFViaMS(1,45,0.4,0.475,0.65,0.775)
% LVxDigEllipBPFViaMS(1,60,0.45,0.55,0.7,0.83)
% LVxDigEllipBPFViaMS(1,75,0.4,0.475,0.65,0.775)
% LVxDigEllipBPFViaMS(1,75,0.1,0.12,0.4,0.5)

15. Write a script that will design a Chebyshev Type-I digital bandpass filter using the impulse invariance method; the script should conform to the call below. Test the script with the given test calls.

function LVxDesignDigCheby1BPFViaImpInv(Ws1,Wp1,...
Wp2,Ws2,Rp,As,Fs)
% Designs a digital bandpass Chebyshev Type I filter using
% impulse invariance. The four bandpass filter band limits
% are Ws1,Wp1,Wp2,and Ws2. The maximum allowable
% passband ripple in positive dB is Rp and the minimum
% stopband attenuation in positive dB is As.
% Test calls:
% LVxDesignDigCheby1BPFViaImpInv(0.2*pi,0.3*pi,...
% 0.6*pi,0.8*pi,0.5,40,0.05)
% LVxDesignDigCheby1BPFViaImpInv(0.3*pi,0.38*pi,...
% 0.6*pi,0.72*pi,0.5,50,0.05)
% LVxDesignDigCheby1BPFViaImpInv(0.2*pi,...
% 0.35*pi,0.6*pi,0.9*pi,1,60,0.1)

16. In this project, we'll modify and use the script

$$[T, F, B] = LVx\_DetnFiltFIRContTone(A, Freq, RorSS, SzWin, OvrLap, AudSig)$$

to identify an interfering steady-state or rising-amplitude sinusoid and remove it from the test signal using a Chebyshev Type-I filter designed automatically in response to the list of candidate interfering frequencies generated through analysis of the spectrogram matrix.

The previous script

LVx_DetnFiltFIRContTone

is modified only in that the filter to be used is a Chebyshev Type-I IIR rather than an FIR. The format of the new script is as follows:

function [ToneFreq,F,B] = LVx_DetnFiltIIRContTone(A,Freq,...
RorSS,SzWin,OvrLap,AudSig,Rp,FltOrd)
% This program evaluates the list of candidate interfering tones
% in the same manner as the script LVx_DetnFiltFIRContTone
% with the exception that filtering is accomplished using a
% Chebyshev Type-I filter (lowpass, notch, or highpass,
% depending on the location of the interfering frequency,
% with maximum passband ripple of Rp dB and order
% FltOrd). The first six input arguments and the output
% arguments of this script are identical to those
% of LVx_DetnFiltFIRContTone.
% Test calls:
% [T,F,B] = LVx_DetnFiltIIRContTone(0.015,200,0,512,1,1,1,2)
% [T,F,B] = LVx_DetnFiltIIRContTone(0.01,210,0,512,1,1,1,2)
% [T,F,B] = LVx_DetnFiltIIRContTone(0.01,200,0,512,1,2,1,2)
% [T,F,B] = LVx_DetnFiltIIRContTone(0.1,200,0,512,1,3,1,2)
% [T,F,B] = LVx_DetnFiltIIRContTone(0.01,500,0,512,1,1,1,2)
% [T,F,B] = LVx_DetnFiltIIRContTone(0.01,500,0,512,1,2,1,2)
% [T,F,B] = LVx_DetnFiltIIRContTone(0.08,500,0,512,1,3,1,2)
%
% [T,F,B] = LVx_DetnFiltIIRContTone(0.025,200,1,512,1,1,1,2)
% [T,F,B] = LVx_DetnFiltIIRContTone(0.02,200,1,512,1,2,1,2)
% [T,F,B] = LVx_DetnFiltIIRContTone(0.1,200,1,512,1,3,1,2)
% [T,F,B] = LVx_DetnFiltIIRContTone(0.02,500,1,512,1,1,1,2)
% [T,F,B] = LVx_DetnFiltIIRContTone(0.018,500,1,512,1,2,1,2)
% [T,F,B] = LVx_DetnFiltIIRContTone(0.1,500,1,512,1,3,1,2)
% [T,F,B] = LVx_DetnFiltIIRContTone(0.07,150,0,512,1,3,1,2)
%
% [T,F,B] = LVx_DetnFiltIIRContTone(0.005,1000,0,512,0,1,1,2)
% [T,F,B] = LVx_DetnFiltIIRContTone(0.008,1000,0,512,1,1,1,2)
% [T,F,B] = LVx_DetnFiltIIRContTone(0.005,1000,0,512,1,1,1,2)

The call

[T,F,B] = LVx_DetnFiltIIRContTone(0.015,120,0,512,0,1,1,2)

results in Figs. 3.40 and 3.41, in addition to the basic figures found in *LVx_DetectContTone* and *LVx_DetnFiltFIRContTone*.

The call

$$[T,F,B] = LVx\_DetnFiltIIRContTone(0.015,500,0,512,0,1,1,2)$$

results in Figs. 3.42 and 3.43.

16. Write the m-code for the script

$$LVx Prony(Imp, NumA, NumB)$$

as described in the text and as defined below, and test it with the given sample calls.

```
% function [b,a] = LVxProny(Imp,NumA,NumB)
% Uses Prony's Method to model an impulse response as an
% IIR having z-transform with NumB numerator coefficients
% b and NumA denominator coefficients a.
% Sample calls:
% [b, a] = LVxProny([(0.9).^( 0:1:50 )],2,2)
% [b,a] = cheby1(2,0.5,0.5), Imp = filter(b,a,[1,zeros(1,75)]);...
% [b,a]=LVxProny(Imp,3,3)
% [b,a] = cheby1(2,0.5,0.5), Imp = filter(b,a,[1,zeros(1,25)]);,,,
% [b,a]=LVxProny(Imp,3,3)
% [b,a] = cheby1(2,0.5,0.5), Imp = filter(b,a,[1,zeros(1,10)]);...
% [b,a]=LVxProny(Imp,3,3)
% [b,a] = cheby1(2,0.5,0.5), Imp = filter(b,a,[1,zeros(1,4)]);...
% [b,a]=LVxProny(Imp,3,3)
% [b,a] = cheby1(2,0.5,0.5), Imp = filter(b,a,[1,zeros(1,3)]);...
% [b,a]=LVxProny(Imp,3,3)
```

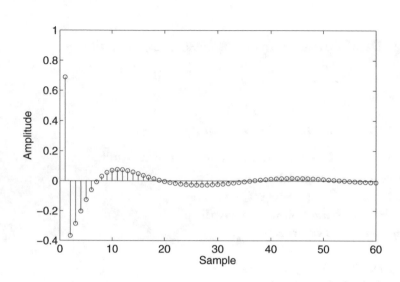

Figure 3.40: Impulse response (truncated) of HPF automatically designed by the script to eliminate an interfering tone at 120 Hz.

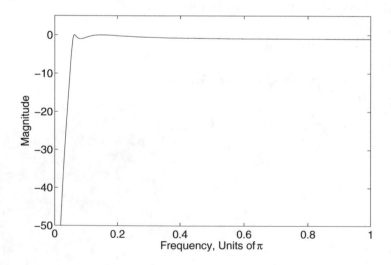

Figure 3.41: Frequency response of the 120 Hz HPF impulse response shown in the previous figure.

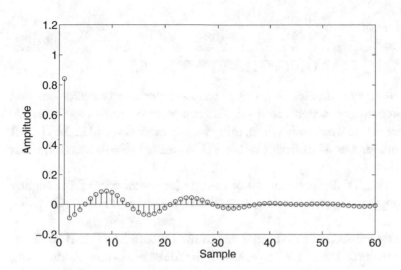

Figure 3.42: Impulse response (truncated) of a notch filter automatically designed by the script to eliminate an interfering tone at 500 Hz.

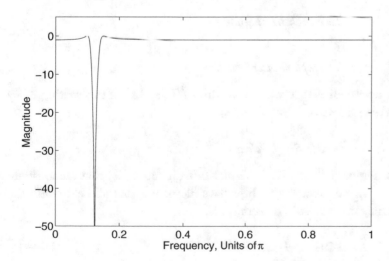

Figure 3.43: Frequency response of the notch filter designed to eliminate a 500 Hz tone.

APPENDIX A

Software for Use with this Book

A.1   FILE TYPES AND NAMING CONVENTIONS

The text of this book describes many computer programs or scripts to perform computations that illustrate the various signal processing concepts discussed. The computer language used is usually referred to as **m-code** (or as an **m-file** when in file form, using the file extension **.m**) in MATLAB -related literature or discussions, and as **MathScript** in LabVIEW-related discussions (the terms are used interchangeably in this book).

The MATLAB and LabVIEW implementations of m-code (or MathScript) differ slightly (Lab-VIEW's version, for example, at the time of this writing, does not implement Handle Graphics, as does MATLAB).

The book contains mostly scripts that have been tested to run on both MATLAB and Lab-VIEW; these scripts all begin with the letters **LV** and end with the file extension **.m**. Additionally, scripts starting with the letters **LVx** are intended as exercises, in which the student is guided to write the code (the author's solutions, however, are included in the software package and will run when properly called on the Command Line).

Examples are:

LVPlotUnitImpSeq.m

LVxComplexPowerSeries.m

There are also a small number m-files that will run only in MATLAB, as of this writing. They all begin with the letters *ML*. An example is:

ML_SinglePole.m

Additionally, there are a number of LabVIEW Virtual Instruments (VIs) that demonstrate various concepts or properties of signal processing. These have file names that all begin with the letters *Demo* and end with the file extension *.vi*. An example is:

DemoComplexPowerSeriesVI.vi

Finally, there are several sound files that are used with some of the exercises; these are all in the .wav format. An example is:

drwatsonSR4K.wav

A.2   DOWNLOADING THE SOFTWARE

All of the software files needed for use with the book are available for download from the following website:

http://www.morganclaypool.com/page/isen

The entire software package should be stored in a single folder on the user's computer, and the full file name of the folder must be placed on the MATLAB or LabVIEW search path in accordance with the instructions provided by the respective software vendor.

A.3   USING THE SOFTWARE

In MATLAB, once the folder containing the software has been placed on the search path, any script may be run by typing the name (without the file extension, but with any necessary input arguments in parentheses) on the Command Line in the Command Window and pressing *Return*.

In LabVIEW, from the Getting Started window, select MathScript Window from the Tools menu, and the Command Window will be found in the lower left area of the MathScript window. Enter the script name (without the file extension, but with any necessary input arguments in parentheses) in the Command Window and press *Return*. This procedure is essentially the same as that for MATLAB.

Example calls that can be entered on the Command Line and run are

LVAliasing(100,1002)

LV_FFT(8,0)

In the text, many "live" calls (like those just given) are found. All such calls are in boldface as shown in the examples above. When using an electronic version of the book, these can usually be copied and pasted into the Command Line of MATLAB or LabVIEW and run by pressing *Return*. When using a printed copy of the book, it is possible to manually type function calls into the Command Line, but there is also one stored m-file (in the software download package) per chapter that contains clean copies of all the m-code examples from the text of the respective chapter, suitable for copying (these files are described more completely below in the section entitled "Multi-line m-code examples"). There are two general types of m-code examples, single-line function calls and multi-line code examples. Both are discussed immediately below.

A.4   SINGLE-LINE FUNCTION CALLS

The first type of script mentioned above, a named- or defined-function script, is one in which a function is defined; it starts with the word "function" and includes the following, from left to right:

any output arguments, the equal sign, the function name, and, in parentheses immediately following the function name, any input arguments. The function name must always be identical to the file name. An example of a named-function script, is as follows:

> **function nY = LVMakePeriodicSeq(y,N)**
> **% LVMakePeriodicSeq([1 2 3 4],2)**
> **y = y(:); nY = y*([ones(1,N)]); nY = nY(:)';**

For the above function, the output argument is *nY*, the function name is *LVMakePeriodicSeq*, and there are two input arguments, *y* and *N*, that must be supplied with a call to run the function. Functions, in order to be used, must be stored in file form, i.e., as an m-file. The function *LVMakePeriodicSeq* can have only one corresponding file name, which is

LVMakePeriodicSeq.m

In the code above, note that the function definition is on the first line, and an example call that you can paste into the Command Line (after removing or simply not copying the percent sign at the beginning of the line, which marks the line as a comment line) and run by pressing *Return*. Thus you would enter on the Command Line the following, and then press *Return*:

nY = LVMakePeriodicSeq([1,2,3,4],2)

In the above call, note that the output argument has been included; if you do not want the value (or array of values) for the output variable to be displayed in the Command window, place a semicolon after the call:

nY = LVMakePeriodicSeq([1,2,3,4],2);

If you want to see, for example, just the first five values of the output, use the above code to suppress the entire output, and then call for just the number of values that you want to see in the Command window:

nY = LVMakePeriodicSeq([1,2,3,4],2);nY1to5 = nY(1:5)

The result from making the above call is

nY1to5 = [1,2,3,4,1]

A.5   MULTI-LINE M-CODE EXAMPLES

There are also entire multi-line scripts in the text that appear in boldface type; they may or may not include named-functions, but there is always m-code with them in excess of that needed to make a simple function-call. An example might be

N=54; k = 9; x = cos(2*pi*k*(0:1:N-1)/N);
LVFreqResp(x, 500)

Note in the above that there is a named-function (*LVFreqResp*) call, preceded by m-code to define an input argument for the call. Code segments like that above must either be (completely) copied and pasted into the Command Line or manually typed into the Command Line. Copy-and-Paste can often be successfully done directly from a pdf version of the book. This often results in problems (described below), and accordingly, an m-file containing clean copies of most m-code programs from each chapter is supplied with the software package. Most of the calls or multi-line m-code examples from the text that the reader might wish to make are present in m-files such as

McodeVolume1Chapter4.m

McodeVolume2Chapter3.m

and so forth. There is one such file for each chapter of each book, except Chapter 1 of Volume I, which has no m-code examples.

A.6   HOW TO SUCCESSFULLY COPY-AND-PASTE M-CODE

M-code can usually be copied directly from a pdf copy of the book, although a number of minor, easily correctible problems can occur. Two characters, the symbol for raising a number to a power, the circumflex ^, and the symbol for vector or matrix transposition, the apostrophe or single quote mark ', are coded for pdf using characters that are non-native to m-code. While these two symbols may look proper in the pdf file, when pasted into the Command line of MATLAB, they will appear in red.

A first way to avoid this copying problem, of course, is simply to use the m-code files described above to copy m-code from. This is probably the most time-efficient method of handling the problem—avoiding it altogether.

A second method to correct the circumflex-and-single-quote problem, if you do want to copy directly from a pdf document, is to simply replace each offending character (circumflex or single quote) by the equivalent one typed from your keyboard. When proper, all such characters will appear in black rather than red in MATLAB. In LabVIEW, the pre-compiler will throw an error at the first such character and cite the line and column number of its location. Simply manually retype/replace each offending character. Since there are usually no more than a few such characters, manually replacing/retyping is quite fast.

Yet a third way (which is usually more time consuming than the two methods described above) to correct the circumflex and apostrophe is to use the function *Reformat*, which is supplied with the software package. To use it, all the copied code from the pdf file is reformatted by hand into one horizontal line, with delimiters (commas or semicolons) inserted (if not already present) where lines have been concatenated. For example, suppose you had copied

```
n = 0:1:4;
y = 2.^n
stem(n,y);
```

where the circumflex is the improper version for use in m-code. We reformat the code into one horizontal line, adding a comma after the second line (a semicolon suppresses computed output on the Command line, while a comma does not), and enclose this string with apostrophes (or single quotes), as shown, where *Reformat* corrects the improper circumflex and *eval* evaluates the string, i.e., runs the code.

$$\text{eval(Reformat('n=0:1:4;y=2.^n;stem(n,y)'))}$$

Occasionally, when copying from the pdf file, essential blank spaces are dropped in the copied result and it is necessary to identify where this has happened and restore the missing space. A common place that this occurs is after a "for" statement. The usual error returned when trying to run the code is that there is an unmatched "end" statement or that there has been an improper use of the reserved word "end". This is caused by the elision of the "for" statement with the ensuing code and is easily corrected by restoring the missing blank space after the "for" statement. Note that the function *Reformat* does not correct for this problem.

A.7   LEARNING TO USE M-CODE

While the intent of this book is to teach the principles of digital signal processing rather than the use of m-code per se, the reader will find that the scripts provided in the text and with the software package will provide many examples of m-code programming starting with simple scripts and functions early in the book to much more involved scripts later in the book, including scripts for use with MATLAB that make extensive use of MATLAB objects such as push buttons, edit boxes, drop-down menus, etc.

Thus the complexity of the m-code examples and exercises progresses throughout the book apace with the complexity of signal processing concepts presented. It is unlikely that the reader or student will find it necessary to separately or explicitly study m-code programming, although it will occasionally be necessary and useful to use the online MATLAB or LabVIEW help files for explanation of the use of, or call syntax of, various built-in functions.

A.8   WHAT YOU NEED WITH MATLAB AND LABVIEW

If you are using a professional edition of MATLAB, you'll need the Signal Processing Toolbox in addition to MATLAB itself. The student version of MATLAB includes the Signal Processing Toolbox.

If you are using either the student or professional edition of LabVIEW, it must be at least Version 8.5 to run the m-files that accompany this book, and to properly run the VIs you'll need the Control Design Toolkit or the newer Control Design and Simulation Module (which is included in the student version of LabVIEW).

Vector/Matrix Operations in M-Code

B.1   ROW AND COLUMN VECTORS

Vectors may be either row vectors or column vectors. A typical row vector in m-code might be [3 -1 2 4] or [3,-1,2, 4] (elements in a row can be separated by either commas or spaces), and would appear conventionally as a row:

$$\begin{bmatrix} 3 & -1 & 2 & 4 \end{bmatrix}$$

The same, notated as a column vector, would be [3,-1,2,4]' or [3; -1; 2; 4], where the semicolon sets off different matrix rows:

$$\begin{bmatrix} 3 \\ -1 \\ 2 \\ 4 \end{bmatrix}$$

Notated on paper, a row vector has one row and plural columns, whereas a column vector appears as one column with plural rows.

B.2   VECTOR PRODUCTS

B.2.1   INNER PRODUCT

A row vector and a column vector of the same length as the row vector can be multiplied two different ways, to yield two different results. With the row vector on the left and the column vector on the right,

$$\begin{bmatrix} 1 & 2 & 3 & 4 \end{bmatrix} \begin{bmatrix} 4 \\ 3 \\ 2 \\ 1 \end{bmatrix} = 20$$

corresponding elements of each vector are multiplied, and all products are summed. This is called the **Inner Product**. A typical computation would be

$$[1, 2, 3, 4] * [4; 3; 2; 1] = (1)(4) + (2)(3) + (3)(2) + (4)(1) = 20$$

B.2.2   OUTER PRODUCT

An **Outer Product** results from placing the column vector on the left, and the row vector on the right:

$$\begin{bmatrix} 4 \\ 3 \\ 2 \\ 1 \end{bmatrix} \begin{bmatrix} 1 & 2 & 3 & 4 \end{bmatrix} = \begin{bmatrix} 4 & 8 & 12 & 16 \\ 3 & 6 & 9 & 12 \\ 2 & 4 & 6 & 8 \\ 1 & 2 & 3 & 4 \end{bmatrix}$$

The computation is as follows:

$$[4; 3; 2; 1] * [1, 2, 3, 4] = [4, 3, 2, 1; 8, 6, 4, 2; 12, 9, 6, 3; 16, 12, 8, 4]$$

Note that each column in the output matrix is the column of the input column vector, scaled by a column (which is a single value) in the row vector.

B.2.3   PRODUCT OF CORRESPONDING VALUES

Two vectors (or matrices) of exactly the same dimensions may be multiplied on a value-by-value basis by using the notation " .* " (a period followed by an asterisk). Thus two row vectors or two column vectors can be multiplied in this way, and result in a row vector or column vector having the same length as the original two vectors. For example, for two column vectors, we get

$$[1; 2; 3]. * [4; 5; 6] = [4; 10; 18]$$

and for row vectors, we get

$$[1, 2, 3]. * [4, 5, 6] = [4, 10, 18]$$

B.3   MATRIX MULTIPLIED BY A VECTOR OR MATRIX

An m by n matrix, meaning a matrix having m rows and n columns, can be multiplied from the right by an n by 1 column vector, which results in an m by 1 column vector. For example,

$$[1, 2, 1; 2, 1, 2] * [4; 5; 6] = [20; 25]$$

Or, written in standard matrix form:

$$\begin{bmatrix} 1 & 2 & 1 \\ 2 & 1 & 2 \end{bmatrix} \begin{bmatrix} 4 \\ 5 \\ 6 \end{bmatrix} = \begin{bmatrix} 4 \\ 8 \end{bmatrix} + \begin{bmatrix} 10 \\ 5 \end{bmatrix} + \begin{bmatrix} 6 \\ 12 \end{bmatrix} = \begin{bmatrix} 20 \\ 25 \end{bmatrix} \qquad (B.1)$$

An m by n matrix can be multiplied from the right by an n by p matrix, resulting in an m by p matrix. Each column of the n by p matrix operates on the m by n matrix as shown in (B.1), and creates another column in the n by p output matrix.

B.4   MATRIX INVERSE AND PSEUDO-INVERSE

Consider the matrix equation

$$\begin{bmatrix} 1 & 4 \\ 3 & -2 \end{bmatrix} \begin{bmatrix} a \\ b \end{bmatrix} = \begin{bmatrix} -2 \\ 3 \end{bmatrix} \tag{B.2}$$

which can be symbolically represented as

$$[M][V] = [C]$$

or simply

$$MV = C$$

and which represents the system of two equations

$$a + 4b = -2$$

$$3a - 2b = 3$$

that can be solved, for example, by scaling the upper equation by -3 and adding to the lower equation

$$-3a - 12b = 6$$

$$3a - 2b = 3$$

which yields

$$-14b = 9$$

or

$$b = -9/14$$

and

$$a = 4/7$$

The inverse of a matrix M is defined as M^{-1} such that

$$MM^{-1} = I$$

where I is called the Identity matrix and consists of all zeros except for the left-to-right downsloping diagonal which is all ones. The Identity matrix is so-called since, for example,

$$\begin{bmatrix} 1 & 0 \\ 0 & 1 \end{bmatrix} \begin{bmatrix} a \\ b \end{bmatrix} = \begin{bmatrix} a \\ b \end{bmatrix}$$

The pseudo-inverse M^{-1} of a matrix M is defined such that

$$M^{-1} M = I$$

System B.2 can also be solved by use of the pseudo-inverse

$$\left[M^{-1} \right] [M] [V] = \left[M^{-1} \right] [C]$$

which yields

$$[I][V] = V = \left[M^{-1} \right] [C]$$

In concrete terms, we get

$$\left[M^{-1} \right] \begin{bmatrix} 1 & 4 \\ 3 & -2 \end{bmatrix} \begin{bmatrix} a \\ b \end{bmatrix} = \left[M^{-1} \right] \begin{bmatrix} -2 \\ 3 \end{bmatrix} \tag{B.3}$$

which reduces to

$$\begin{bmatrix} a \\ b \end{bmatrix} = \left[M^{-1} \right] \begin{bmatrix} -2 \\ 3 \end{bmatrix}$$

We can compute the pseudo-inverse M^{-1} and the final solution using the built-in MathScript function *pinv*:

```
M = [1,4;3,-2];
P = pinv(M)
ans = P*[-2;3]
```

which yields

$$P = \begin{bmatrix} 0.1429 & 0.2857 \\ 0.2143 & -0.0714 \end{bmatrix}$$

and therefore

$$\begin{bmatrix} a \\ b \end{bmatrix} = \begin{bmatrix} 0.1429 & 0.2857 \\ 0.2143 & -0.0714 \end{bmatrix} \begin{bmatrix} -2 \\ 3 \end{bmatrix}$$

which yields $a = 0.5714$ and $b = -0.6429$ which are the same as 4/7 and -9/14, respectively. A unique solution is possible only when M is square and all rows linearly independent.(a linearly independent row cannot be formed or does not consist solely of a linear combination of other rows in the matrix).

APPENDIX C

FIR Frequency Sampling Design Formulas

In the formulas below, L represents FIR length in samples, and the values A_k are frequency sample amplitudes as described in Section 2.6 of this volume.

C.1   WHOLE-CYCLE MODE FILTER FORMULAS

C.1.1   ODD LENGTH, SYMMETRIC (TYPE I)

The design formula for an odd length linear phase FIR (whole-cycle mode) is

$$h[n] = \frac{1}{L}\left[A_0 + \sum_{k=1}^{M} 2A_k \cos(2\pi(n-M)k/L) \right] \tag{C.1}$$

where the index n for the impulse response $h[n]$ runs from 0 to L - 1, and $M = (L-1)/2$.

C.1.2   EVEN LENGTH, SYMMETRIC (TYPE II)

In the case of even length, $M = (L-1)/2$, is a non-integer (an odd multiple of 1/2) and the design formula is

$$h[n] = \frac{1}{L}\left[A_0 + \sum_{k=1}^{L/2-1} 2A_k \cos(2\pi(n-M)k/L) \right]$$

where the index n for the impulse response $h[n]$ runs from 0 to $L-1$. Note that even length filters need the 1/2 sample offset in the sample index n in order to make the cosine components symmetrical about their midpoint.

C.1.3   ODD LENGTH, ANTI-SYMMETRIC (TYPE III)

$$h[n] = \frac{1}{L}\sum_{k=1}^{M} 2A_k \sin(2\pi(M-n)k/L)$$

where L is the filter length, $M = (L-1)/2$, and n runs from 0 to $L-1$. Note that k may not equal zero here, or rather, the sine of zero is identically zero, so no DC (frequency 0) correlator can be generated using sine waves as the basis, which explains why the Type III and IV filters do not work as lowpass filters.

C.1.4 EVEN LENGTH, SYMMETRIC (TYPE IV)

$$h[n] = \frac{1}{L} \left[\sum_{k=1}^{L/2-1} 2A_k \sin(2\pi(M-n)k/L) + A_{L/2}\sin(\pi(M-n)) \right]$$

where L is the filter length, $M = (L-1)/2$, and n runs from 0 to $L-1$.

C.2 HALF-CYCLE MODE FILTERS

These filters are built from odd-multiples of half-cycles of cosines or sines.

C.2.1 ODD LENGTH, SYMMETRIC (TYPE I)

$$h[n] = \frac{1}{L} \left[\sum_{k=0}^{M-1} 2A_k \cos(2\pi(n-M)(k+\tfrac{1}{2})/L) + A_M\cos(\pi(n-M)) \right]$$

where $M = (L-1)/2$, and n runs from 0 to $L-1$.

C.2.2 EVEN LENGTH, SYMMETRIC (TYPE II)

$$h[n] = \frac{1}{L} \left[\sum_{k=0}^{N/2-1} 2A_k \cos(2\pi(n-M)(k+\tfrac{1}{2})/L) \right]$$

where $M = (L-1)/2$, and n runs from 0 to $L-1$.

C.2.3 ODD LENGTH, ANTI-SYMMETRIC (TYPE III)

$$h[n] = \frac{1}{L} \left[\sum_{k=0}^{M-1} 2A_k \sin(2\pi(M-n)(k+\tfrac{1}{2})/L) \right]$$

where $M = (L-1)/2$, and n runs from 0 to $L-1$.

C.2.4 EVEN LENGTH, ANTI-SYMMETRIC (TYPE IV)

$$h[n] = \frac{1}{L} \left[\sum_{k=0}^{L/2-1} 2A_k \sin(2\pi(M-n)(k+\tfrac{1}{2})/L) \right]$$

where $M = (L-1)/2$, and n runs from 0 to $L-1$.

C.3 REFERENCES

[1] T. W. Parks and C. S. Burrus, *Digital Filter Design*, John Wiley & Sons, New York, 1987.

Biography

Forester W. Isen received the B.S. degree from the U. S. Naval Academy in 1971 (majoring in mathematics with additional studies in physics and engineering), and the M. Eng. (EE) degree from the University of Louisville in 1978, and spent a career dealing with intellectual property matters at a government agency working in, and then supervising, the examination and consideration of both technical and legal matters pertaining to the granting of patent rights in the areas of electronic music, horology, and audio and telephony systems (AM and FM stereo, hearing aids, transducer structures, Active Noise Cancellation, PA Systems, Equalizers, Echo Cancellers, etc.). Since retiring from government service at the end of 2004, he worked during 2005 as a consultant in database development, and then subsequently spent several years writing the four-volume series DSP for MATLAB and LabVIEW, calling on his many years of practical experience to create a book on DSP fundamentals that includes not only traditional mathematics and exercises, but "first principle" views and explanations that promote the reader's understanding of the material from an intuitive and practical point of view, as well as a large number of accompanying scripts (to be run on MATLAB or LabVIEW) designed to bring to life the many signal processing concepts discussed in the series.

Printed in the United States
by Baker & Taylor Publisher Services